MATTERS OF CARE

CARY WOLFE, *Series Editor*

(*continued on page 267*)

Matters of Care

Speculative Ethics in
More Than Human Worlds

MARÍA PUIG DE LA BELLACASA

posthumanities **41**

University of Minnesota Press
Minneapolis • London

An earlier version of chapter 1 was published as "Matters of Care in Technoscience: Assembling Neglected Things," *Social Studies of Science* 41, no. 1 (2011): 85–106. An earlier version of chapter 2 was published as "Nothing Comes without Its World: Thinking with Care," *The Sociological Review* 60, no. 2 (2012): 197–216. An earlier version of chapter 3 was published as "Touching Technologies, Touching Visions: The Reclaiming of Sensorial Experience and the Politics of Speculative Thinking," *Subjectivity* 28 (2009): 297–315. An earlier version of chapter 4 was published as "Ethical Doings in Naturecultures," *Ethics, Place, and Environment: A Journal of Philosophy and Geography* 13, no. 2 (2011): 151–69. An earlier version of chapter 5 was published as "Making Time for Soil: Technoscientific Futurity and the Pace of Care," *Social Studies of Science* 45, no. 5 (2015): 691–716.

Published by the University of Minnesota Press
111 Third Avenue South, Suite 290
Minneapolis, MN 55401-2520
http://www.upress.umn.edu

The University of Minnesota is an equal-opportunity educator and employer.

Library of Congress Cataloging-in-Publication Data
Names: Puig de la Bellacasa, María, author.
Title: Matters of care : speculative ethics in more than human worlds / María Puig de la Bellacasa.
Description: Minneapolis : University of Minnesota Press, 2017. |
Series: Posthumanities ; 41 | Includes bibliographical references and index.
Identifiers: LCCN 2016015356 | ISBN 978-1-5179-0064-9 (hc) |
ISBN 978-1-5179-0065-6 (pb)
Subjects: LCSH: Caring. | Ethics.
Classification: LCC BJ1475 .P85 2017 | DDC 177/.7—dc23
LC record available at https://lccn.loc.gov/2016015356

For
COYOTE

Contents

The Disruptive Thought of Care

Care, caring, carer. Burdened words, contested words. And yet so common in everyday life, as if care was evident, beyond particular expertise or knowledge. Most of us need care, feel care, are cared for, or encounter care, in one way or another. Care is omnipresent, even through the effects of its absence. Like a longing emanating from the troubles of neglect, it passes within, across, throughout things. Its lack undoes, allows unraveling. To care can feel good; it can also feel awful. It can do good; it can oppress. Its essential character to humans and countless living beings makes it all the most susceptible to convey control. But what is care? Is it an affection? A moral obligation? Work? A burden? A joy? Something we can learn or practice? Something we just do? Care means all these things and different things to different people, in different situations. So while ways of caring can be identified, researched, and understood concretely and empirically, care remains ambivalent in significance and ontology.

Embracing these ambivalent grounds, not without tentativeness, this book invites a speculative exploration of the significance of care for thinking and living in more than human worlds. I choose this phrasing among other existing ways of naming posthumanist constituencies because it speaks in one breath of nonhumans and other than humans such as things, objects, other animals, living beings, organisms, physical forces, spiritual entities, and humans.[1] Encompassing this ontological scope is vital as it has become indisputable, if it ever wasn't, that in times binding techno-sciences with naturecultures, the livelihoods and fates of so many kinds and entities on this planet are unavoidably entangled. Certainly this term

1

remains unsatisfying, for its abstract unspecificity, and for the moral under-
tones that invite us to "transcend" the human for something "more than."
It also still starts from a human center, then to reach "beyond." However, it
works well enough as the uncertain terrain for the delicate task of broad-
ening consideration of the lives involved in caring agencies, still mostly
thought as something that human people do. Care is a human trouble, but
this does not make of care a human-only matter. Affirming the absurdity
of disentangling human and nonhuman relations of care and the ethicalities involved requires decentering human agencies, as well as remaining
close to the predicaments and inheritances of situated human doings.

I feel grounded and supported in this effort by a crowd of thinkers,
researchers, and activists who might not endorse this endeavor but to
which it is nonetheless indebted. It feels reductive to try to account for the
richness of work on care in this introduction. But at least a note has to
be made for readers who might be unfamiliar with the inheritances of a
project that seeks once again to affirm care despite and because of its
ambivalent significance. Certainly any notion that care is a warm pleasant
affection or a moralistic feel-good attitude is complicated by feminist
research and theories about care. Since Carol Gilligan's famous and controversial *In a Different Voice* rooted the origins of a caring ethical subjectivity in the mother–child relation (Gilligan 1982), the discussion of the moral
and political value of the work of care, as well as the thorough inquiry
by feminist sociologies into the different labors that involve and make
care, has been expanded and challenged from a range of perspectives that
go well beyond activities traditionally and socially identified as women's
work. The well-known discussions on the "ethics of care" are just a small
part of conversations that collect an extended range of interlocutors not
necessarily aware of each other's voice. The evidences of care have been
challenged for more than thirty years now through nursing studies, sociologies of medicine, health and illness, and ethics and philosophy, as well
as political thought. With or beyond the ethics of care, practices and principles of care have been explored critically in the domains of critical psy-
chology (Noddings 1984), political theory (Tronto 1993), justice (Engster
2009), citizenship (Kershaw 2005; Sevenhuijsen 1998), migration and labor
studies (Boris and Rhacel 2010), care in business ethics and economics

(Gatzia 2011), scientific choices for development (Nair 2001), in sociologies and anthropologies of health work and sciences (Latimer 2000; Mol 2008; Mol, Moser, and Pols 2010; Lappé, forthcoming), disability studies and activism (Sánchez Criado, Rodríguez-Giralt, and Mencaroni, 2016), care in accountability procedures (Jerak-Zuiderent 2015), food politics (Abbots, Lavis, and Attala 2015), as an ethics for animal rights (Donovan and Adams 2010), and in farming practices (Singleton and Law 2013)—not to speak of research rooted in grassroots activism (Precarias a la Deriva 2004; Barbagallo and Federici 2012), social and health work, and policy (Hankivsky 2004). Closer to the specific trajectories of this book, care is also explored as a significant notion to appreciate affective and ethico-political dimensions in practices of knowledge and scientific work (Rose 1983, 1994; Despret 2004; Muller 2012; Suzuki 2015; Perez-Bustos 2014) and as a politics in technoscience (Martin, Myers, and Viseu 2015) with a vital significance for ecology (Curtin 1993) and human–nonhuman relations in naturecultural worlds (Haraway 2011; Van Dooren 2014; Kirksey, 2015).

The list could go on and continues to expand; even attempting a sample feels reductive. All these engagements with care make specific contributions to the understanding and meanings of care, revealing how caring implicates different relationalities, issues, and practices in different settings. These investigations might not all agree with each other—nor should they have to—about what care means or involves. Nevertheless, specific inquiries into actualizations of care have also contributed and coexist with a theoretical discussion of care as a "generic" doing of ontological significance, as a "species activity" with ethical, social, political, and cultural implications. For Joan Tronto and Bernice Fischer, this includes *everything that we do* to maintain, continue and repair "our world" so that we can live in it as well as possible. That world includes our bodies, our selves, and our environment, *all of which we seek to interweave in a complex, life-sustaining web* (Tronto 1993, 103, emphasis added).

I pause a moment on this much-quoted generic definition of care because it emerges at several moments in this book. It became, inadvertently, not only a conception I kept coming back to, like a reassuring refrain allowing to touch ground along the meanders of a speculative journey, but

also an enticement to probe further into the meanings of care for thinking
and living with more than human worlds. What is included in "our" world?
And why should relations of care be articulated from there?[2]

But before coming to these questions, I want to unpack in this introduction some of the reasons why this definition of care became a point of
departure. Joan Tronto dedicated her book *Moral Boundaries*—still one of
the most influential works on care and a landmark piece of feminist political philosophy and ethics—to unpack the political significance of care. In
this spirit, her generic definition of care emphasizes an extended notion of
the agencies encompassed by care: "everything we do to." Among these
doings it both discriminates and keeps tightly together the "maintenance"
aspects of care—what is traditionally referred to as "care work"—and the
sense of an ethics and politics of care, the pursuit of a "good" life, expressed
in an affectively charged "as well as possible." Tronto also articulated the
dimensions that join to generate an "integrated" act of care: the affective
and ethical dispositions involved in concern, worry, and taking responsibility for other's well-being, such as "caring about" and "taking care of,"
need to be supported by material practices—traditionally understood as
maintenance or concrete work involved in actualizing care, such as "care
giving" and "care receiving" (Tronto 1993, 105–8; Sevenhuijsen 1998). The
distinction does not separate these modes of agency. What it allows us
to emphasize is that a politics of care engages much more than a moral
stance; it involves affective, ethical, and hands-on agencies of practical and
material consequence. Another critical dimension of this generic conception is the accent on care as vital in interweaving a web of life, expressing
a key theme in feminist ethics, an emphasis on interconnection and interdependency in spite of the aversion to "dependency" in modern industrialized societies that still give prime value to individual agency. While this
field is often focused on unpacking the specificity of "dependency work"—
necessary when we are unable to take care of ourselves (Kittay 1999; Kittay
and Feder 2002)—it also suggests interdependency as the ontological state
in which humans and countless other beings unavoidably live. This doesn't
mean that dependency is an absolute value in all situations—as critics in
disability studies as well as struggles for independent living expose well
(Kröger 2009)—nor that dependency and independency are antithetic.

Care is not about fusion; it can be about the right distance (see chapter 3). It also doesn't mean that *to care* should be a moral obligation in all situations, practices, or decisions. Virginia Woolf spoke compellingly of the power of cultivating indifference as a form of quiet revolt, the disruptive power of choosing not to care about what we are enjoined to (Woolf 1996). It does mean, however, that for interdependent beings in more than human entanglements, there has to be some form of care going on somewhere in the substrate of their world for living to be possible. And this is one way of looking at relations, not the only one.

Care as a concrete work of maintenance, with ethical and affective implications, and as a vital politics in interdependent worlds is an important conception that this book inherits from. These three dimensions of care—labor/work, affect/affections, ethics/politics—are not necessarily equally distributed in all relational situations, nor do they sit together without tensions and contradictions, but they are held together and sometimes challenge each other in the idea of care I am thinking with in this book. Instead of focusing on the affective sides of care (love and affection, for instance), or on care as work of maintenance, staying with the unsolved tensions and relations between these dimensions helps us to keep close to the ambivalent terrains of care. There are situations when care work involves a removal of the affective—we ask, then, why would a paid care worker have to involve affection in her work? This is crucial because we have to consider how care can turn into moral pressure for workers who might rightfully want to preserve their affective engagement from exploitations of waged labor. But if maintenance does not involve some affective involvement—I care for, I worry (or I am summoned to even if I don't want to)—is it still about care? In contrast, one can also love intensely without committing to the "work of love," without involvement in the sometimes tedious maintenance of a relation. That we ask such questions reveals that affectivity—not necessarily positive—is part of situations of care, as oppressive burden, as joy, as boredom. Staying with these tensions exposes that vital maintenance is not sufficient for a relation to involve care, but that without maintenance work, affectivity does not make it up to care and keeps it closer to a moral intention, to a disposition to "care about," without putting in the work to "care for" (Tronto 1993). The same applies to ethical and political questions raised by care, such as outrage and condemnations about its absence, or about controlling policies that regulate what is considered legitimate care. While it might seem that care does not necessarily involve an ethics or a politics, there seems to be an inherent positioning that happens through engagements with caring. As Anne Marie Mol emphasizes in her elucidation of a logic of care in doctoring practices: "Articulating 'good care' . . . is an intervention" rather than a factual evaluation or judgement of practice (Mol 2008, 84). In this book I thus explore care as intrinsically involving an ethical and political intervention that affects also those who are researching care. Because speaking of "good care"—or of as-well-as-possible care—is never neutral. Because the work of care can be done within and for worlds that we might find objectionable. For what is care given? These interrogative oscillations expose how Tronto's definition of care—by keeping the tensions between care as maintenance doings and work, affective engagement, and ethico-political involvement—opens a terrain for exploring, in situation, the subtle thought of care, by reading these dimensions through each other.

Moreover, the generic character of this definition of care is also particularly engaging for a speculative project. First, because it exposes the existential domains of care as something open-ended—everything we do. Second, because it points to how the "ethics" in an ethics of care cannot be about a realm of normative moral obligations but rather about thick, impure, involvement in a world where the question of how to care needs to be posed. That is, it makes of ethics a hands-on, ongoing process of re-creation of "as well as possible" relations and therefore one that requires a speculative opening about what a possible involves. And thus the thinking in this book is moved by a generic appeal of care that makes it unthinkable as something abstracted from its situatedness. So while this book is not a sociological or ethnographic inquiry into a specific domain of agencies of care, by engaging speculatively with the meanings of thinking and living with care I hope to contribute to an enrichment of its meanings in a way that invites others to consider care—or its absence—as a parameter of existence with significance for their own terrains. Yet far from a general treaty on care, this book attempts to generate its own situatedness in the interplay between the generic and the specific.[3] Each chapter presents and opens questions around care starting from different angles, questions, and problems, specific matters of care so that the generic does not resolve in a closed theory. Looking at problems and research terrains as matters of care can then also become a speculative research question as these researchers propose: "The question, then, is not 'how can we care more?' but instead to ask what happens to our work when we pay attention to moments where the question of 'how to care?' is insistent but not easily answerable. In this way, we use care as an analytic or provocation, more than a predetermined set of affective practices" (Atkinson-Graham et al. 2015).

Another thinking motif in the book, the connection of "speculative" and "ethics," might also need some clarification, though it is meant to be quite simple. A discussion of ethicality and obligation runs across the book, but the approach does not aim at systematic ethical theorizing (I invite readers intrigued by these "ethics" to start reading from chapter 4). The journey is about ethics because it raises questions about the meanings of care for *as well as possible* worlds but is also marked by a trajectory that brought me to think about care first as a political commitment. The speculative then connects to a feminist tradition for which this mode of thought about the possible is about provoking political and ethical imagination in the present.[4] But the ethical discussions in this book are also speculative because they try to avoid defining a normative framework for how to make the "as well as possible" as they displace the meanings of care in terrains where they could disturb the meanings of an established "well." Affirming the speculative as a general orientation, of course, somehow presupposes a critical approach to the present. Why would one want other possible worlds if nothing was wrong with this one? Therefore a hesitant search for what it means to think critically and speculatively is woven throughout the book. But affirming the speculative in ethics invokes an indecisive critical approach, one that doesn't seek refuge in the stances it takes, aware and appreciative of the vulnerability of any position on the "as well as possible."

Finally, there is a more "empirical" reason why the significance of *critical speculative* thought became part of this exploration of care. Care is not only ontologically but politically ambivalent. We learn from feminist approaches that it is not a notion to embrace innocently. Thought and work on care still has to confront the tricky grounds of essentializing women's experiences (Hoagland 1991) and the persistent idea that care refers, or should refer, to a somehow wholesome or unpolluted pleasant ethical realm. Delving into feminist work on the topic invites us to become substantially involved with care as a living terrain that seems to need to be constantly reclaimed from idealized meanings, from the constructed evidence that, for instance, associates care with a form of unmediated work of love accomplished by idealized carers. Contemporary reengagements with care are keeping this outlook when they both engage to continue fostering care as well as warn against an overoptimistic view on its practice when they prompt us to continue "unsettling" care (Murphy 2015), or as Aryn Martin, Natasha Myers, and Ana Viseu put it, prolonging Donna Haraway's call, to "stay with the trouble" (Haraway 2016) in the way we engage in caring. Those involved in thinking about critically complex terrains and struggles of care for more than forty years might not feel there is anything new in these concerns. And yet, we cannot settle on the idea that these discussions are behind us. A colleague told me recently about her reaction to the organization of a symposium on care and political economy: "Care? Why should I care about care?" then pausing, she recounted how as she was hearing herself think this way, she thought: "probably because everything tells me not to care about care." Beyond the obvious point that what matters to a generation will not continue per se to matter into the future, care is so vital to the fabric of life that it remains an ongoing matter of struggle and a terrain of constant normative appropriation.

And it is true that in spite of a tradition that has made of care an essential feature of transformative thinking, politics, and alternative forms of organizing, care is also a commonplace topic in everyday moralizations, especially in the West, or Global North. A hegemonic revival that sees care valued beyond traditional domains could be reinforced by a present permeated by worries about the unraveling of life from all possible crisis fronts—environment, economy, values. And while the sense of emergency translates into constant anxiety, into the expectation of a catastrophic event (Beuret 2015), a less broadcasted background violence slowly destroys more fundamentally the tissue of everyday existence for living beings at all scales (Nixon 2011). This sense of crisis and the need to care more is stressed by the perspective of a few, albeit powerful, ontological loci that had benefited from a relative sense of "security" marketed as the norm, while "the rest" of the world, at home and beyond, could carelessly be left in a state of exception (Brown et al. 2012). If only we *all* could care! Really? And what would that mean?

Calls for caring are everywhere, from the marketing of green products, by which companies compete to show how much *they* care, to the purchase of recycled items, by which consumers show that *we* care (Goodman 2013; Goodman and Boyd 2011). More profound and preoccupying beyond this moral marketing gloss is how neoliberal governance has made of caring for *the self* a pervasive order of individualized biopolitical morality. People are summoned to care for everything but, foremost, for "our" selves, our lifestyle, our bodies, our physical and mental fitness, or that of "our" families, reducing care to its most "parochial" caricature (Tronto 1993). Those considered as traditional carers—women generally—or as typical professional carers—nurses and other marginalized unpaid or low-paid care workers—are constantly moralized for not caring enough, or not caring "anymore," or for having "lost" some "natural" capacity to care.[5] And it is not only the present uses of the notion but also past enactments of care that could be reexamined. Michelle Murphy shows in her research on the women's health movement, which many of us still cherish as a model of reappropriation of the means of reproduction, how projects driven by a notion of care can serve colonizing projects (Murphy 2015). Care can be instrumentalized at a global political level too. In a strong critique of humanitarian campaigns in migration contexts that enact "transnational regimes of care," Miriam Ticktin shows how in the name of a universalist idea of relieving suffering—and what Tronto would have called paternalistic care—these actions are rather perpetuating inequalities and preventing collective change that could make a difference for migrant lives (Ticktin 2011).

While there is nothing new in the entanglements of care with hegemonic regimes—one can only think of all the ways in which the caring mother is historically enthroned as much as confined and her caring body enlisted for the nation—different situated configurations will require critical engagements from those who have been trying to have care valued as absolutely vital in the weaving of existence and who will not simply rejoice at the overuse of the word. Some might prefer to let go of a notion far too prevalent in established moral orders. And yet there are many reasons to treat the reductive appropriation of care in the contexts of the ethical ideologies of the Global North with attention rather than scorn. In such a world, a range of different understandings and appropriations of care are made possible and need to be problematized. These will add layers of complexity to feminist visions of care and allow us to avoid reductionist simplifications of the good and evils of care. The picture on the ground is always more fuzzy, and contemporary engagements with care in new terrains continue to show this. Ethnographies of care show how absurd it is to disentangle care from its messy worldliness. Anne Marie Mol shows the ways in which a logic of care and a logic of choice are in constant friction in medical practices (Mol 2008), Sonja Zerak Juiderent exposes forms of situated care that persist within logics of accountability procedures that are seemingly "subjecting care" to abstract norms (Jerak-Zuiderent 2015), Kris Kortright elicits practices of care at the heart of the laboring of a new Green-GMO revolution (Kortright 2013), Wakana Suzuki exposes how an ethos of caring attention is explicitly mobilized as part of a discipline in the laboratories of biomedicine she has observed (Suzuki 2015). Expanding the sites and constituencies in which we think with care contributes new modes of attention and problematics. So rather than give up on care because it is enlisted in purposes we might deplore, we need to have its meanings debated, unpacked, and reenacted in an implicated way that responds to this present.

Care is too important to give it up to the reductions of hegemonic ethics. Thinking *in* the world involves acknowledging our own involvements in perpetuating dominant values rather than retreating to the sheltered position of an enlightened outsider who knows better. Can thinking be connected if it pretends to be outside of worlds we want to see transformed, even those we would rather not endorse? My intention in this book is not to stage a detached confrontation with mainstream notions of care, nor even to deconstruct or police sentiments about care as something that warms hearts and relations—such as the expectation that care brings good—but rather, to propose modes to contribute to its re-articulation, re-conception, and "re-enactment" (King 2012).[6] This requires taking part in the ongoing, complex, and elusive task of *reclaiming* care not from its impurities but rather from tendencies to smooth out its asperities—whether by idealizing or denigrating it. Certainly to reclaim often means to reappropriate a toxic terrain, a field of domination, making it again capable of nurturing; the transformative seeds we wish to sow. It also evokes the work of recuperating previously neglected grounds. But more important for the approach to care in this book, reclaiming requires acknowledging poisons in the grounds that we inhabit rather than expecting to find an outside alternative, untouched by trouble, a final balance—or a definitive critique. Reclaiming is here all but about purging and "cleaning" a notion; rather, it involves considering purist ambitions—whether these are moral, political, or affective—as the utmost poisonous. Reclaiming as political work points to an ongoing effort within existing conditions without accepting them as given. It implies not shying away from what is important to us only because it has been "recuperated" by power, or by hype. This effort is for me an attempt to prolong a style of thought learned through feminist efforts to foster solidarities between divergent feminist positions without erasing unresolvable tensions (Puig de la Bellacasa 2013; 2014). While feminist materialist analysis of care exposed society's reliance on care—its importance—it also revealed the intricacies of the work made by carers, showing how relations of dependency care can be cruel as much as loving, unpacking what is actually *done* in different situations under the blanket category of care. Reclaiming care is to keep it grounded in practical engagements with situated material conditions that often expose tensions. Rather than engaging with discussions of care in their different specific configurations and specialist knowledge, this book inherits from ongoing conversations the assumption that the meanings and situated relevance of care cannot be taken for granted. Assuming how intrinsic care is to the everyday fabric of troubled worlds, I try not to pin down care to one of its ontological dimensions—affective, practical, ethico-political—and embrace its ambivalent character. Inspired by Leigh Star's insurgent approach to exclusions, to the silences and violences implied in the evidences of naming, pinning down, and classifying (see, for instance, Leigh Star 1991), I resist categorizing care but rather seek to emphasize its potential to disrupt the status quo and to unhinge some of the moral rigidities of ethical questioning.

My approach to displacing care by involving it in issues and debates in which it has not frequently been addressed is positioned within contemporary critical reorientations but not so much by engaging with a critique of care. I am thinking care itself as a critically disruptive doing that can open to "as well as possible" reconfigurations engaged with troubled presents. Thus critically but speculatively, this book stays with the transformative potential care, despite and because of hegemonic ethics, of its current commodification, despite and because care's unescapable importance makes it vulnerable to become a powerful vehicle of normative moralization. Staying with care's potential to disrupt thus is not (only) about making visible neglected activities we want to see more "valued"—for instance, as "productive" activities with an economic worth that should be recognized. It requires engaging with situated recognitions of care's importance that operate displacements in established hierarchies of value and understanding how divergent modes of valuing care coexist and co-make each other in non-innocent ways. So if this book can contribute to the meaningfulness of care, I hope it will be by adding layers to perceptions of care, by avoiding the smoothing out of its disruptive potential. It is standing on these ambivalent grounds, but firm in an obstinate conviction of the existential and ontological significance of care, that I attempt to expand thinking on care by moving the investigation of its meanings into a mostly unchartered terrain: the meanings of care for knowing and thinking with more than human worlds in technoscience and naturecultures.

Displacing Care

A broad understanding of posthumanist thought includes work that, increasingly in the past twenty years, has questioned the boundaries that pretend to define the human realm (against the other than human as well as otherized humans), to sanction humanity's separate and exceptional character and, purposely or not, to sanction the subjection of *everything else* to this purported superiority. The frontiers blurred through these ways of thinking and the sociomaterial moves that impel them are now commonly known: between nature and culture, society and science, technology and organism, humans and other living forms. The thinking at stake is transdisciplinary to the core, involving a wide range of perspectives and

methodologies in the social sciences and humanities that form also relatively new fields: science and technology studies, animal studies, posthumanist philosophy and ethics, environmental humanities. The cultural,
political, and ethical challenges are colossal and the search for alternatives
ongoing. In this book, I modestly try to contribute to these efforts.

It could be simply said that the thoughts presented here basically displace usual meanings of care just because they are put to work around
problems that disrupt the classic boundaries that feminist politics of care
have mostly worked with in order to claim the significance of care for
social worlds. But this task involves making also some nonobvious moves.
The first one, in Part I, is a transfer of meaning that carries the triptych of
care as "ethics-work-affect" into the terrain of the politics of knowledge,
into the implications of thinking with care. John Dewey, playing with the
semantic affinities of care and "mind," of caring and mindful, said beautifully that 'mind' denotes every mode and variety of interest in, and concern for things. . . . In respect to situations, events, objects, persons and
groups . . . it signifies memory . . . attention . . . purpose. . . . Mind is care"
(Dewey 1958, 263).[7] This is an appealing notion, by which the relational
character of thought (as mind) is rendered as care. But thinking and knowing are often not caring, not even mindful, nor is caring through knowing
and thinking an unproblematic endeavor. With this awareness at heart,
Part I engages with discussions of knowledge politics in thought that
engages beyond human agencies. Part II attempts to displace questions
that have been mostly asked about care as something human subjects do—
and some surrogate persons such as nonhumans deemed capable of intentional agency and emotion. What does caring mean when we go about
thinking and living interdependently with beings other than human, in
"more than human" worlds? Can we think of care as an obligation that traverses the nature/culture bifurcation without simply reinstating the binaries and moralism of anthropocentric ethics? How can engaging with care
help us to think of ethical "obligations" in human-decentered cosmologies?

In order to start exposing how such questions might be addressed, it is
time now to go into more detail about how these two displacements unfold
in this book. Part I, *Knowledge Politics*, starts by setting the initial motivations for expanding the ethico-political meanings of care. Across the three

chapters in Part I runs a preoccupation with knowledge politics in techno-science. Thus the book starts by engaging with discussions in science and technology studies (STS) that address the "more than human worlds" of sociotechnical assemblages and objects as lively politically charged "things." These three chapters are marked by the context of my own encounter with the notion of care through feminist work that is not typically identi-fied with discussions of care. Early Marxist-feminist materialist thinking from the late 1980s, often known as "standpoint feminist theory," explored the possibility of a feminist epistemology and rooted the hopeful prospect of alternative ways of knowing in the materiality of women's and other mar-ginalized people's everyday experiences. A discussion, reenactment, and partial prolongation of these discussions unfolds in the first three chap-ters. It was in the relational confrontation with the everyday maintain-ing of life that other forms of knowing were posited as possible, one that could deeply understand the importance of material mediations against the abstractions of "masculine" thought established on detachment from these devalued activities (Hartsock 1983; Smith 1987; Collins 1986; Harding 1991). Care here is featured as a part of those labors that mediated with the material world, in particular domestic and family care labors, tra-ditionally the realm, and existential confinement, of women, especially from underprivileged class and racial backgrounds. One particular text associated with such discussions marked me. *Hand, Brain, and Heart: A Feminist Epistemology for the Natural Sciences,* by the British feminist soci-ologist of science Hilary Rose, explored the political significance of car-ing to subvert the industrial-military-scientific complex (1983; 1994). She spoke of women's movements such as the Greenham Common Women's Peace Camp against nuclear weapons, which used symbols of care to cre-ate disruption, threading baby socks into wired fences, but also displaced women's identity as caring mothers into a public sphere of direct action against nuclear weapons—and were sadly disqualified as bad mothers for leaving families behind to do so. Rose also spoke of aerospace workers who moved from participating in the manufacture of war technologies to designing socially beneficial technologies. Like others, Rose saw care as grounded in the material conditions of women's reproductive labors, and she associated care with the working dimension of love, but the gist of her project was to bring the obligation of care as a way to contest the dominant ways of knowledge and science production in technoscience. This way, she laid bare the potential of care's generic significance to confront and disrupt the destructive dynamics of scientific knowledge that separates brain and hand, intellect and practice, from the "heart."

It is Rose's insight that initiated me on a path for thinking care as a politics of knowledge at the heart of technoscientific, naturecultural worlds. Hilary Rose's conception is marked as much by radical feminist knowledge politics as by "radical science" movements (Rose and Rose 1976). Like contemporary early sociologies of science, this work understood that sciences and technologies are permeated by politics and ethics to the core rather than, as traditionally conceived, in their application or "use/abuse" by society. But specifically, this is work that sustained a firmly critical attack on the pervasive exclusions and violences inherent to technoscience. Politics in this early type of "radical science studies" of the late 1970s and early 1980s was more than an analytical parameter of sociological inquiry; it encompassed a commitment and positioning of the knowledge *we* produce for "as well as possible" worlds. This political urge contrasts with the predominance of more "neutral" approaches to politics and ethics developed since the bourgeoning of analysis focusing on technoscience that converged in forming the loose interdisciplinary field of Science and Technology Studies (STS). Some have argued that the critique of science simply became "academic" and the commitment to critical intervention faded away (Martin 1993; for a recent, similar argument, see Mirowski 2015). Another translation of this is the qualification of the field as description-oriented—in the wake of the Actor Network Theory of "following" the actors on the ground—rather than normative (i.e., ethically or politically oriented to impose an ideological "should be"). While these critiques are without doubt accurate in some ways, blanket judgments of depoliticization also tend to disregard that a discussion of forms of ethical and political involvement in approaches to technoscience has never been completely closed; it is a recurrent theme, and one that has remained particularly alive in influential feminist work in the field (see, for instance, Mayberry, Subramaniam, and Weasel 2001; Star 1995; Suchman 2007a) and other explicitly positioned approaches (Hess 2007; Winner 1986; see Sismondo 2008 and 2010 for an overview).

And today it is possible to perceive a renewed interest in more "explicit" forms of engagement with the politics of the production of knowledge in technoscience. Here a notion of care has also become a way to name an ethico-political practice and an affective engagement within knowledge production about technoscience. Notions such as a "radicalization" of care and a "politics of care" are addressing the nature of "implication" and "relevance" (Savransky 2014) of intellectual and research work as intervention.[8] Caring also becomes a way of speaking of the critical engagements of knowledge producers beyond the polarized divisions around the meanings of social and politically "useful" research (Metzger 2013; 2014). One of the initial and ongoing motivations in writing this book is partly situated in this collective reenactment of committed knowledge as a form of care. There is renovated enthusiasm in these moves that is relevant not only to a reinvestment in a critique of contemporary technoscience but, more important for this book, to a search to prolong transformative knowledge as it is involved in troubled worlds after the fundamental lesson of contemporary STS: not only that knowledge and science are material-semiotic affairs with strongly political and ethical consequences but that a decentered conception of human agency exposes relations with objects, things, and other than human animals, organisms, and forms as political in their very ontology.

Thinking about and with care is compelling in this context because it offers possibilities for thinking commitment and obligation as nonnormative forms of ethical engagement that could be more attuned to the decentering of human agency and privilege in contemporary thinking of technoscience and naturecultures. But this assumption is just a starting point. Reconnecting a politics of commitment and of ethical obligation with an ontology of more than human worlds without falling back into classic humanist categories of thought requires a speculative effort. It specifically poses the question of the compatibility of distributed agency and decentering the human subject with situated ethical obligations and commitments. That this is a tricky problem is well stated in how Lucy Suchman reminds us that when engaging with technoscientific assemblages "the price in recognizing the agency of artefacts need not be the denial of our own" (Suchman 2007b, 285). In this direction, the discussion of care in this book ultimately relates to how we conceive of a critical or, political, ethos in posthumanist thinking, of an "insurgent posthumanism" (Papadopoulos 2010). Indeed, the reclamation of care in approaches to more than human worlds marked by technoscience is a political project that defies the traditional ethical boundaries that have marked critical thinking. Following the trope of care into an "unexpected country" (Haraway 2011) of blurred boundaries—moral as much as material—requires opening up its possible meanings.

Starting from these discussions, Part I engages with the politics of thinking and knowing in the more than human worlds of technoscience—mostly involving "things" and objects or, broadly speaking, material-semiotic agencies mobilized by science and technology (while Part II attends to relations of caring in more than human living ecologies). It locates the discussions within technoscience, understood basically as a world and time in which scientific knowledge and the material production of technologies are inseparable from sociopolitical processes and imaginaries, including those of commodification. Technoscience as the world where knowledge is inseparable from material worlds—where knowledge is involved in making things matter—is here conceived as literally as possible: material-semiotic agency in the mattering of worlds. As Karen Barad notes, it is through entangled agencies and practices of matter and meaning that technoscientific worlds "come to matter" (Barad 2007). So as the first three chapters unfold, questions about the politics of knowledge in technoscience increasingly delve into ethical concerns raised by our proximity and involvement with the material effects of our thought. Worlds seen through care accentuate a sense of interdependency and involvement. What challenges are posed to critical thinking by increased acute awareness of its material consequences? What happens when thinking about and with others is understood as *living* with them? When the effects of caring, or not, are brought closer? Here, knowledge that fosters caring for neglected things enters in tension between a critical stance against neglect and the fostering of speculative commitment to think how things could be different.

Chapters 1 and 2 engage with these questions through what can be read as a contrast between two close readings of Bruno Latour's and Donna Haraway's work. The thinking, concepts, and research objects, but mostly the knowledge politics of these authors, are terrains to think speculatively about what caring knowledge politics could mean in more than human worlds. Speaking of contrast here is not meant to create opposition but to unpack propositions that both diverge and communicate through connected concerns. Chapter 1, *Assembling Neglected "Things,"* engages with the politics and agency of things in science and technology studies. It is articulated around a commentary and prolongation of Latour's notion of "matters of concern." The chapter makes explicit the notion that gives title to this book, "matters of care," as one that inscribes care in the materiality of more than human things. It inherits from a now well-established tradition that rejects the representation of science, technology, and nature as depoliticized matters of fact, as uncontestable truths. I bring to the center of these interrogations feminist research and thinking in science and technology studies as work that remains crucial in encouraging an ethos of care not only in thinking the processes of construction of socio-technical assemblages but as an ethico-political attitude in the everyday doing of knowledge practices. Latour's naming of matters of fact as "matters of concern" attracts attention to the ethico-political effects of constructivist accounts in science and technology studies as they attempt to make things matter by "re-presenting" things. Concern brings us closer to a notion of care. However, there is a "critical" edge to care that the politics of making things matter as gatherings of concerns tends to neglect. Against this background I explore what it would mean to think matters of fact and socio-technical assemblages as matters of care. Can more awareness regarding concerns favor the promotion of care in contemporary technoscience? Can an affective ethico-political concern such as caring become a thinking pattern when engaging with science and technology? This chapter tries to respond to these questions without seeking a normative answer and by drawing upon feminist knowledge politics and theories of care, as well as commenting on empirical research in the field of science and technology studies that expands and reaffirms the importance and meanings of care. I read this work not so much to develop or discuss the substantiality of their contributions—that is, their research on specific cases of care—but in search of *ways of thinking* that engage care. Positioning for care emerges as an oppositional practice that both creates trouble in the democratic assembly of articulate concerns as well as generates possibility: it reminds us of exclusions and suffering and fosters alternative affective involvements with the becomings of science and technology. Rather than defining moral parameters for these positionings, I ask a speculative question "how to care?" about the ways "things" are constructed, presented, and studied, especially when care seems to be expendable.

Chapter 2, *Thinking with Care,* probes further into imagining how a style of thought can contribute to caring thinking in *living* with other than humans. If bringing care to matter started in the previous chapter as a requisite for knowledge that aims to re-present things, here knowledge is further conceived as embedded in the mattering of worlds. This chapter expands the premise that thinking and knowing are essentially relational processes that require care. Grounded in this relational conception of ontology inspired by Tronto's web of care, I explore "thinking with care" as a thick, noninnocent requisite of collective thinking in interdependent worlds. This speculative exploration of motifs thinking with care unfolds through a reading of Donna Haraway's work, specifically her take on the situated character of knowledge. A notion of thinking with care is articulated through the chapter as a series of concrete moves: *thinking-with,* *dissenting-within,* and *thinking-for.* While weaving Haraway's thinking and writing practices with the trope of care offers a particular understanding of this author's knowledge politics, the task of caring knowledge also emerges as more challenging. Exploring the generic notion of care through a confrontation with the more than human worlds in which "staying with the trouble" appears as the only ethical option for knowledge mattering (Haraway 2016) shows again the potential of care to create trouble in established logics, as well as possibility.

Caring thinking needs to resist an idealized version of knowledge politics. Chapter 3, *Touching Visions,* builds on this understanding by reading caring thinking and knowing as touch. Touch, or the haptic, could be the sensorial universe that better explores the ambivalences of conceiving caring knowledge as an intensification of involvement and proximity. Touch is also the sensorial metaphor that better exposes qualms around the materiality of thinking and its consequential effects: we think, therefore we touch. But this exploration of touch attempts to be in itself an exercise in carefulness about the speculative potentialities of haptic visions. In other words, (my)

efforts to reclaim touch—aka proximal intimate knowing—as a neglected
way of knowing need to resist an idealized version of knowing-touching.
This discussion inherits both from work in feminist knowledge politics
and conceptions of science and technology that problematize epistemological distances—between subjects and objects, knowledge and "the world,"
and science and politics. In this direction, touch expresses a sense of
material-embodied relationality that seemingly eschews abstractions and
detachments that have been associated with dominant epistemologies of
knowledge-as-vision. Touch becomes a metaphor of transformative knowledge at the same time as it intensifies awareness of the imports of speculative
thinking. In other words, the haptic disrupts the prominence of vision as a
metaphor for distant knowing as well the distance of critique, but it also
calls for ethical questioning. What is caring touch in this context? Here,
somehow paradoxically, thinking touch with care troubles the desires for
immanent proximities as susceptible to reproducing the negation of mediations and the nonevidence of ethical reciprocity. The terrain around which
I articulate these arguments is the revaluation of the sense of touch, from
cultural theory to expanding markets of haptic technologies. Instances of
haptic fascination expose not only the potential of thinking with literal and
figural meanings of touch but also the temptations of idealizing materiality. Yet engaging speculatively with experience, knowledge, and technology
as touch allows us to explore a possible transformation of ethos that could be
brought by more careful touching visions and the forms of ethical obligation
they entail. In particular, touch's unique quality of reversibility, that is, the
fact of being touched by what we touch, puts the question of reciprocity at
the heart of thinking and living with care. What's more, the reciprocity of
care is rarely bilateral, the living web of care is not maintained by individuals giving and receiving back again but by a collective disseminated force.
Thus conceived, the complexity of the circulation of care feels even more
all-pervasive when we think of how it is sustained in more than human
worlds. Care is a force distributed across a multiplicity of agencies and
materials and supports our worlds as a thick mesh of relational obligation.

Part II, *Speculative Ethics in Antiecological Times*, moves through this living mesh as it engages with everyday ecologies of sustaining and perpetuating life for their potential to transform entrenched relations to natural

worlds as "resources." If the focus of the first chapters was on technoscience, here care matters pertain to relational webs in *naturecultures*. Though distinguishing technoscience from naturecultures is in many ways meaningless in contemporary political ecologies, other ethico-political and affective questions of care emerge when they involve humans-and-other-species (see Latimer and Miele 2013) as well as other than human (though not human-free) entities such as biophysical energies and elements. While Part I engages with conceptual work, approaching other thinkers' thought and research, as well as cultural phenomena, as its materials, as its matters of care, the two last chapters bring in my own experiential research in two overlapping terrains of ecological ethics: the practices of the permaculture movement and the transformation of human–soil relations around a notion of soil as living. This order manifests an underlying direction in the book that is not intended as a hierarchical prominence that puts concepts and thinkers first and substance second (or that goes from theory to practices). The truth is more biographical, and it exposes an engagement with the concept of care that evolved from being embedded in philosophies of science and the politics of knowledge (affected by my philosophical background and by becoming a scholar in the midst of the explosion of epistemological debate in the early 1990s) to become absorbed in science studies, a field rich in situated ethnographies that enticed me to want to tell stories. Maybe inadvertently my work also followed a "turn to ontology" (Lezaun and Woolgar 2013) that affected the field itself and relegated epistemological interest or, in more generous terms, rematerialized it. So while the thinking in Part II remains conceptual, the interventions are the most "empirically" grounded—as if the need to treat care in situatedness intensified as it became more layered with meaning, and therefore closer to be presented through "terrains." Yet the thinking efforts appear at their most speculative—as if the limits of what I can think with (my) available ethical notions became more acute when confronted with actual emergent relational arrangements that require making care central without reinstating a human center. What does it mean to think of agencies of care in more than human terms? Engaged more substantially and deeply in telling stories around experientially observed and researched terrains makes the complexities of thinking with care even more intricate. In any case, my relation to these worlds remains openly moved by a commitment to treat the emerging issues as matters of care and therefore attempts to tell involved stories, neither theoretical nor descriptive, open to alternative readings, yet situated.

Chapter 4, *Alterbiopolitics,* proposes a speculative approach to a nature-cultural ethics of care as it manifests in the everyday practices promoted by the international ecological movement known as permaculture. I argue that in order to understand the specific contribution of these forms of ethical engagement without reducing them to "back to nature" ideals or a matter of lifestyle ethics requires us to displace traditional understandings of the ethical. Notwithstanding what can be read as a mostly unsophisticated use of concepts familiar to ethical theory, the chapter does draw on postconventional and poststructuralist ethical approaches that have expanded the limits of ethical discussion. These moves allow us to think the ethics involved in the continuation of life, of *bios,* not so much as a matter of individual morality but as a personal-collective mode of engagement in the everyday that is more about the transformation of ethos than about a normative morality. Discussions about ethics in *biopolitics* are here an entry point for the displacement of the ethical from its status as an edification of a higher morality. But to understand the relevance of an ethics as that of permaculture, embedded in the basic aspects of sustaining and fostering life at its most corporeal levels of naturecultural interdependency—biological and physical—we also need to question the focus on the perpetuation of life *as* human. For this I explore ways in which the notion of "ethical obligation" shifts meaning, from ethical commitments arising out of moral principles—such as contracts or promises—to be embedded in vital material forces involved in the constraints of everyday continuation and maintenance of life. Care troubles and opens questions here too. Connecting the practices of permaculture ethics as everyday ecological doings with a feminist notion of care displaces biopolitical moralities, allowing us to envision *alterbiopolitics* as an ethics of collective empowerment that puts caring at the heart of the search of everyday struggles for hopeful flourishing of *all* beings, of *bios* understood as a more than human community.

The final chapter of the book, *Soil Times*, examines contemporary transformations in human–soil relations happening at the interface of scientific conceptions of soils and ecological practices and that are remattering soil from inert, usable substance and resource into a living world of which humans are also part. It is based on a review of the literature of the soil sciences and research on connected domains of production of knowledge around soil, including permaculture. My reading is oriented by the speculative project of looking out for those displacements where a difference is being made in ways to care for the soil. The dominant human relation to soil has been to pace its fertility with production demand. But today public consideration for soils is changing, amid concerns that they have been mistreated and neglected by the productionist drive. Soils are perceived as endangered ecologies in need of urgent care and warnings about their exhaustion are marked by concerns about a gloomy future that prompt us to act *now*. This chapter introduces a new motif in the discussion, that of the temporalities of care. The pace required by ecological relations with soils could be at odds with accelerated, future-oriented responses characteristic of the pace of technoscientific innovation. Here, making time for care time appears as a disruption of anthropocentered temporalities. Contrasted but interconnected temporalities are at work in contemporary conceptions of soil care in scientific research and other domains of soil practice. Alternative practical, ethical and affective ecologies of care are emerging that trouble the traditional direction of progress and the speed of technoscientific, productionist, future-driven interventions. Among these are the current trends in scientific conceptions of soil that depart from a notion of soil as resource and receptacle for crop production to emphasize its status as a living world. In this context, a "foodweb" model of soil ecology has become a symbol of embodied, caring, involvement with soils. Focusing on the temporal experience of soil care at play in these conceptions reveals a diversity of interdependent temporalities of beings and things at the heart of the predominant futuristic timescales of technoscientific expectations.

The chapter opens to the books conclusions, with thoughts on the untimely character of care in the current political economies of productionist, future-oriented technoscience. Making time for care appears as a material effort for speculative ethical commitments in more than human worlds marked by technoscientific and naturecultural relations. I have tried in the book to approach tensions without succumbing to easy oppositions, thickening the meanings of care as a noninnocent but necessary ethos of always situated implications. Reaching the end, the reader will, I hope, see how the generic notion of care with which this journey starts has become extended but also challenged. Its ambivalences deepened without diminishing the urge to keep practices of care within our thinking spectrum when seeking ways of living together as well as possible. From the perspective of human–nonhuman relations in technoscience and naturecultures, unproblematic visions of care—whether as an exploit of higher ethical beings, a marketable productive activity, or even a recuperated morality to reject—would not only be meaningless but could be fatal. We cannot afford to obscure the actual more laborious and situated conditions in which care takes place and by which its agencies circulate in interdependent more than human relational webs. So as the argument in this book progresses, an acute feeling also intensifies: that an ethical reorganization of human–nonhuman relations is vital, but what this means in terms of caring obligations that could enact nonexploitative forms of togetherness cannot be imagined once for all. And so I hope the reader will forgive me if this book opens up more questions than it offers answers.

PART I

Knowledge Politics

It matters what matters we use to think other matters with; it matters what stories we tell to tell other stories with; it matters what knots knot knots, what thoughts think thoughts, what descriptions describe descriptions, what ties tie ties. It matters what stories make worlds, what worlds make stories.

—DONNA HARAWAY, *Staying with the Trouble*

one

Assembling Neglected "Things"

Should we be at war, too, we the scholars, the intellectuals?
Is it really our duty to add deconstruction to destruction?
More iconoclasm to iconoclasm? What has become of the
critical spirit?

—BRUNO LATOUR, *Why Has Critique Run out of Steam?*

This beautiful planet is sore, and bearable living conditions continue to be inaccessible to the many. The joint fortune that immeasurable forms of life share with human technoscience is no longer news. Developing more scientific research and technological solutions continues to be the dominant response to problems globally and locally—whether these concern climate change, economic recessions, food crises, infertility, or access to health care or information. Social and cultural studies of science and technology thrive in this environment. From the most mundane infrastructures of everyday life, dull corners of laboratories, ordinary households and gardens, to the most arcane and techno-hyped spaces of posthuman consumerism, our world has become a research field for investigating networks and ecologies by means of constructivist philosophical approaches and empirical investigations of emerging ontological politics. Research and thinking proliferate on the multiple ways sciences and technologies contribute to disrupt the boundaries between nature and culture, science and society, matter and thought. In such a context, knowledge politics seem to belong to an old-fashioned story, a once upon a time where subjects were subjects and objects were objects, and epistemology the obstinate elephant in the room. A time when the politics of knowledge appeared to be so important that they become ontological, mutating into new materialist approaches that fused knowers with worlds—while knowledge, science, and technology were being consecrated as the driver of political economies—the world as human innovation. Knowledge is not anymore considered a discrete human affair that filters an objective world out there; it is embedded in the ongoing remaking of the world. In this world of imploded frontiers, there is no way to think sentimentally about purportedly pre-technoscientific pasts and no way to think epistemologically straight. But as blurred boundaries deepen entanglements and interdependencies, the ethico-political demand persists and maybe intensifies for elucidating how different configurations of knowledge practices are consequential, contributing to specific rearrangements. Even more than before, knowledge as relating—while thinking, researching, storytelling, wording, accounting—matters in the mattering of worlds.

This context is a given for the project of this book, the terrain that prompted the urge to prolong discussions of care to think speculatively about the persistence of knowledge politics and ethics in more than human worlds. Motivated by the view that care can open new ways of thinking, this chapter and the next ask what it means to encourage an ethos of care when engaging with sociotechnical relationalities of things human and nonhuman that defy the traditional ethical boundaries that have marked critical work. This inquiry takes ground on what can be learned from both empirical studies and critical thinking on actual practices of care and its ethics, to ask a generic, hopefully generative, question: How can an ethico-political concern such as caring affect the involvement of those who set themselves to observe and represent technoscientific agencies, things, and entities in ways that do not reobjectify them? Can care count in this context as more than promoting a responsible maintenance of technologies in naturecultures? And if yes, is it just a moral value added to the thinking of things and sociomaterial assemblages? Does caring knowledge involve an epistemological or/and an ontological move? These are questions that invite us to explore an idea of care that goes beyond moral disposition or a well-intentioned attitude when considering its everyday significance for knowledge formations in technoscientific worlds, that is, in knowledge economies that make it difficult to claim any innocent or outsider position of *observation*.

Drawing upon feminist thinking is helpful in this effort. In feminist discussions as well as in activism, the politics of caring remain at the heart of concerns with exclusions and critiques of power dynamics in stratified worlds. Considering care as a struggle makes of it an ethico-political issue well more problematic than it could initially seem to be. With this awareness in mind and heart, I want to discuss ways in which care can count for engagement with "things" from the perspective of critical interventions in technoscience. Discussions about the politics of things in science and technology studies (STS) are a good point of departure for this book's journey for various reasons. First, given the focus of this interdisciplinary field on thinking together societal and other than human dimensions of science and technology. Second, because feminist work has marked this endeavor with a commitment toward alternative politics of knowledge. Third, because questions regarding the ethico-political implications of STS as a field have been present throughout its formation and development.

Indeed, more generally speaking, the question of knowledge politics is at the heart of science and technology studies, and not just as an "externalist" problem about how politics might affect knowledge and science production. It is an intrinsic technical element to social studies of science and technology to be established on the idea that sciences and technologies are not simply used or misused by sociopolitical interests *after* the hardware job is stabilized in aseptic "neutral" labs. This understanding was at the heart of seminal studies that continue to initiate students in this field (see Shapin and Schaffer 1985; Collins and Pinch 1993; Latour 1987). Following this early constructivist understanding, it became difficult to detach the meaning-producing technologies of the field—methods, theories, and the stories we tell in our witnessing acts as Donna Haraway put it (1997a)— from their sociopolitical aspects and effects. Early interrogations such as Langdon Winner's widely relayed inquiry, "Do Artifacts Have Politics?" (Winner 1986)—which have multiplied in a range of explorations about "how" they have politics rather than "if"—couldn't be just a matter of producing more *accurate* representations of technology by including politics in accounts and cartographies of sociotechnical, naturecultural assemblages.

Rather, those questions pertain also to the politics of our modes of thought and research ethos, which in turn will affect the politics the thinkers of things attribute to objects and nonhumans. From this perspective, every *Dingpolitik*—one of Bruno Latour's telling names for the politics of things (Latour 2005a)—denotes a *thinkpolitics*. Ways of knowing, theories and concepts, what Shapin and Shaffer called the "literary technologies" embedded in material technologies, have ethico-political and affective effects on the perception and refiguration of matters of fact and sociotechnical assemblages—that is, on their material-semiotic existences (Haraway 1991c; Barad 2007). In other words, ways of studying and representing things have world-making effects. Constructivist approaches to science and nature, no matter how descriptive, are actively involved in redoing worlds.

In prolonging these inheritances of constructivist thought, this chapter explores how care can be part of accounts of science's matters of fact and of sociotechnological assemblages. For this I propose a notion of "matters of care" crafted in discussion with problems stirred up by Bruno Latour's idea of "matters of concern" and the knowledge politics underpinning it. I read Latour's move to rename matters of fact as matters of concern as responding to aesthetic, ethico-political, and affective issues faced by constructivist thinking and its particular form of criticism of things. Not only does Latour's notion represent a particularly influential way of conceiving knowledge politics in technoscience, it also introduces the need to care in a particular way. What this conversation with Latour reveals is that the implications of care are thicker than the politics turning around matters of (public) concern might allow thinking. Involving a feminist vision of care in the politics of things both encourages and problematizes the possibility of translating ethico-political caring into ways of thinking with nonhumans.

The Weariness of Critical Constructivism

Latour's notion of matters of concern critically prolongs the early insight of sociologies of science and technology that scientific and technological assemblages are not just objects but knots of social and political interests and therefore "socially constructed" rather than existing objectively as an expression of the laws of the natural world. This vision gained in subtlety with moves in sociopolitical approaches to science and technology, as constructivism moved from being "social" to "ontological," opening a range of perspectives on the possibilities of "ontological politics" (Mol 1999; Papadopoulos 2011; 2014b). Mediations of agency and materiality no longer appear as mastered or directed by human/social subjects but as co-enacted by nonhumans. This conception affects the way we thing about the role of humans, culture and "the social." It is not so much that "social" interests are added to nonhuman worlds by acting upon the scientifically driven course of technological development. Human intervention does not disappear, but agency is distributed. Interests and other affectively animated forces—such as concern and care—are decentered and distributed in fields of meaning-making materialities: from being located in the intentionality of human subjectivity, they become understood as intimately entangled in the ongoing material remaking of the world. It is the ethico-political implications of accounts dedicated to unpacking these intricate agencies in more than human worlds that is well represented by the rebaptism of matters of fact into "matters of concern" by Bruno Latour (Latour 2004b; 2005b). The notion became popular as a renaming that could help to emphasize engaged ethico-political responsiveness in technoscience in an integrated way, that is, within the very life of things rather than through normative added values.

The notion of matters of concern (hereafter MoC) is relatively recent, but the concerns that animate it are not. MoC makes a difference in three sets of problems familiar to philosophical discussions about the politics of science and technology studies in general, and of constructivism in particular. In the first place, MoC prolongs awareness regarding the liveliness of things in continuity with conceptual efforts aimed at disobjectifying scientific matters of fact (Latour 1993; 1999). Latour's work is rich in diplomatic bridging efforts, trying to convince sociologists and humanists that nonhumans have a "soul" and scientists, technologists, and engineers that their facts and artifacts are embodied sociality (Latour 1996a). A first naming that tried to do this was his praise of the "hybrid." Coming back to *We have never been modern*, we can recall the refreshing immersion into the Middle Kingdom, a world of epistemologically puzzling yet ontologically robust hybrids—from global geopolitical entities such as the hole in the ozone layer to prelaboratory devices such as Robert Boyle's legendary air pump. These, Latour argued, had been mistreated as "objects" by the philosophies pledged to the "Modern Constitution"—an arrangement of binary purifications that cuts through the complex human–nonhuman *mediations* that hybrids make happen (and that make hybrids happen) splitting apart their naturecultural, realconstructed, sosialscientific, discursivematerials modes of existence (Latour 1993). Another bad habit of the modern ethos, of the impulse to crack things open, is a gusto for purist dissection, coupled with the dismissive othering of those who do not dissect (e.g., fetishists or premoderns) and, eventually, with the *reduction* of the objectified part of the binary to the other (e.g., technology as object of humans, or vice versa). "Matter of fact" appeared as a poor epistemological category born to this modern tradition that reduces the rich recalcitrant reality of proliferating entities. But while modern thinking kept being utterly wrong about what makes the world go round, in the Middle Kingdom hybrids thrived nonchalant about misled philosophical binaries: the world of mediations is what *always* was—we had never been (really) modern.

Ten years after this influential intervention, Latour continued praising sociologies and anthropologies of science that formed the field of science and technology studies for having understood the realities of the Middle Kingdom and for their continued effort to look for better ways of presenting things differently, in a nonmodern way, a nonhumanist way, that is, of putting into practice a nonobjectifying aesthetics. Part of this approach involves that "when agencies are introduced, they are never presented simply as matters of fact, but always as *matters of concern*, with their mode of fabrication and their stabilizing mechanisms clearly visible" (Latour 2004b, 246, emphasis added). MoC provided a new conceptual tool for this well-explored task: the restaging of things as lively. This aesthetics helps us to resist to what Alfred North Whitehead called the "bifurcation of nature," which splits feelings, meanings, and the like, from the hardcore facts (Latour 2005d, 12; Whitehead 1920). Latour called upon Whitehead to reinstate the diagnosis of a recalcitrant problem: we remain trapped in binary oppositions, in the perception that in order to account for phenomena we need to bridge a gap between two worlds. And even though bridging, we still tend to give one side the power to *know*, and even *do*, the other: nature explains society (or vice versa) (Latour 2005d, 5–6). Following Isabelle Stengers (2000), Latour argued that these bifurcations/gaps/splits between natural facts and social questions preside over the somewhat misled enterprise of calling "social" our constructivism.[1] We need new styles of thinking, new notions to name what we are thinking, to heal from the manic drive to dissect the togetherness that we perceive. Whitehead's suggestion to avoid this bifurcation is that natural philosophy "might not pick and choose," because "*everything* perceived is in nature"—the molecules of the scientists, the meanings of poets (Latour 2005d, 12, emphasis added; Whitehead 1920, 28–29). The Latourian translation became: everything perceived is in the "thing."

The *everything* in the "thing" is read here through the potential significance of the word as "gathering" (a Heideggerian pick, but without much of its Heideggerianism), thought together with other meanings aimed at naming the "many" that makes a "thing"—such as "society" (inspired by Gabriel Tarde's sociology and again Whitehead's metaphysics), a "collective," an "assembly," or an "association." Thus here its renaming as "thing" aims to convey a more lively perception, understanding and *restaging*, of the misnamed objectified matter of fact: aesthetics is politics. Thing, aka gathering, makes patent the internal diversity of "matters of fact," the blurred boundaries of its collective existence, as well as the mediations that make possible for it to hold together as well as constantly build new associations. A thing, conceived as such, is then both construction *and* reality. And if "things" are matters of concern, it is also because they are gathering a collective that forms around a common concern. To be able to think things as such, Latour argues for a new sense of "empirical philosophy" diverging from "flat" empiricist epistemology,[2] one that would place us in the flow of this moving experience. Instead of bridging worlds, we can "drift" in what Whitehead called the "passage of nature" (a more poetic view of Latour's Middle Kingdom), in the dense troubled waters of "*what is given into experience*" (Latour 2005d, 4). The letting go of the controlling power of causal and binary explanation comes with an immersion in the messy world of concerns. Being in the things we plunge into unsettled gatherings; rather than observe them from a bridge, we inhabit the realm of more than human politics.

And so this way of restaging matters of fact has significance for a second familiar theme: the inclusion of things in politics. Following Noortje Marres's call to put "issues" at the heart of politics (Marres 2007), Latour affirmed: "A thing is, in one sense, an object out there and, in another sense, an *issue* very much *in* there, at any rate, a *gathering* . . . the same word *thing* designates matters of fact and matters of concern" (Latour 2005d, 233). In a technical characterization of MoC, the notion appears as "another powerful descriptive tool" in the project to enliven depoliticized things (232) that traverses a search for legitimate "representative" accounts—politically speaking—of nonhuman agency in the networks. In *We Have Never Been Modern* and other interventions, Latour had already celebrated how the anthropological approach to science and technology and nonhuman worlds more generally helped objects become "free citizens" by exhibiting them as "mediators—that is, actors endowed with the capacity to translate what they transport, to redefine it, redeploy it, and also to betray it" (Latour 1993, 81). These agencies were invisible to "human-centered" politics that excluded them and saw them as mere objects—threatening or serviceable to human politics. The target of this critique can be identified as a humanist morality, traditionally oblivious to how scientific matters of fact and technical things "gather," to how they can transform the composition of a world. Instead, Latour argued, thing-oriented politics give them a political "voice." They ask in a more democratic fashion, "How many are we?" in order to include in this "we" the often misrepresented nonhumans, as full participants in public life (Latour 2004a).

MoC appears as yet another name for what sociologies, histories, and anthropology of science contributed to our understanding: that matters of fact and technological assemblages had always been worlds of entangled concerns. MoC is then another tool for the political task of representing things, for the aesthetical move of restaging them. So what did this naming add to the discussion? It was deemed necessary because in spite of the early insights and the recalcitrance of things to reductionisms of all sides, in spite of all the work of actor-network theory and science and technology studies in bringing forward the "reality" of mediations, Latour felt that we were not yet done. MoC aims both to reaffirm and expand awareness about STS's specific aesthetic contribution: the *mise en scène* of the actual imbroglios and "envelopes" that make possible the work of scientists (Latour 2005d). And yet MoC is not just a new name for an old issue. It responds to an added problem: what I call the weariness of critical constructivism. With the introduction of concerns an affectively charged question seems to come into the picture as a way to foster a new style of thinking *in* the things: Haven't we, through the process of opening up things to expose their modes of fabrication, ended up dissecting, disarticulating, lessening their reality?

We could simply say that the notion of MoC translates the political life of things in a language compatible with the changing terminologies of contemporary majoritarian democracies, today dealing with "issues" of "public concern" (and I'll come back to this problem later). But it also goes beyond this. Emphasizing concern stresses the troubled and unsettled ways, the more or less subtle ethical, political, and affective tremors, by which a gathering/thing/issue is constructed and holds together. For me, though the problems MoC encompassed were well known, the introduction of this notion indicated a subtle, yet meaningful, displacement. By contrast with "interest"—a previously prevalent notion in the staging of forces, desires, and the politics sustaining the "fabrication" and "stabilization" of matters of fact—concern alters the affective charge of the thinking and presentation of things with connotations of trouble, worry, and care. The question Latour frames as a "style" is also a problem of knowledge politics: how we present things matters. Replacing interests by concerns as the force of political claims and their inclusion alters in a significant way the material-semiotic perception of things: interests are something that the inheritors of agonistic modern politics have learned to approach suspiciously—or that we are supposed to jealously preserve when they are our "own." Concerns, in turn, call upon our ability to *respect* each other's issues, including those pertaining to nonhuman's lives, if we are to build a common world.

And that respect is also at the heart of a third crucial impulse that I identify at the heart of Latour's proposal for thinking things as MoC: the disempowering effects of constructivism when it concedes too much to "critique" and ends up turning the insight that "facts are constructed" into "disbelief" (Latour 2004b). It was in a humorous and emphatic contribution to *Critical Inquiry* in which the notion of MoC was first proposed. Here Latour urged critical thinkers to treat "matters of fact" as "matters of concern," appealing to a sense of self-protection of our "own" concerns: Would you really appreciate your concerns being reduced, deconstructed, or dismantled? (240). Affirming that matters of fact are matters of concern encourages awareness of the vulnerability of the facts and things we set out to study and criticize. One major symptom of critical excess is for Latour the abuse of notions of "power," used as causal explanations "coming out of the deep dark below" to undermine what others, generally other scientists, present as facts (229). These worries refer to particularly pernicious gestures he qualifies as "critical barbarianism" to which critical thinkers are likely to succumb. Another is antifetishism (Latour 1996b), the utmost disrespectful critical barbaric gesture (akin to "iconoclasm"), by which the insight that "facts are constructed" turns into "disbelief." These explanatory strategies all attempt to debunk the "real," purportedly concealed by fetishes, artifices, beliefs, ideologies, discourses, social structures, or any other terms that causal reasoning might invoke "powerful explanations" underlying the constructions that critique makes its mission to disarticulate.

Dislike regarding critical descriptions that stress power and domination as key social forces that make science and technology—a "lust for power" (Latour 2005b, 85)—is a constant in Latour's thought.[3] Latour has also argued that these are technically inadequate for accounts based on actor-network theory (which typically shouldn't add "ready-made" explanations to the cartography of how actors and networks deploy). But what I find more interesting in the introduction of MoC is how it stresses further the ethico-political and affective effects of critical intervention—not only on things, facts, and the world, but *on those who set out to research them*. This sentiment is well staged with the funny figuration of a tired (social) constructivist who has learned a lesson: today a tragicomically demoted "Zeus of critique," who knows how things *really* work but reigns in a desert alone, loved by none as he has criticized and deconstructed everything. As such, his locomotive has "run out of steam" (Latour 2004b, 239). This vision is consistent with his earlier thought-provoking characterization of the modern spirit: those who *believe* that *others* believe—others that are of course wrong, as they believe things that "we" now know are not true (Latour 1993; see also Stengers 1997). This polemic style of thought works through progressively producing radical or revolutionary breaks with surmounted beliefs that are pushed into an obscure past by new, more enlightened, knowledge (Latour 1993, 72). Characterized as such, as a particular ethos of inquiry, "critique" becomes a transversal heritage that affects all who descend from the modern scientific enterprise (Stengers 2008) rather than being the appanage of a specific academic field (of critical theory). But here the critical constructivist appears exhausted and tormented, suspecting that it could have contributed to the ongoing dismantling of the world. Not any more an outsider, nor even bridging, but immersed in the troubled flow. It is this mood that I characterize as the weariness of critical constructivism, provoked by worry about the effects and contributions of constructivist visions of STS at the heart of a technoscientific culture that keeps producing a fair amount of fright.

Latour adopts the mood of a concerned critic in search of reliable, renewed trust in the reality of matters of fact, still good humored but not as assured as before, somewhat distressed by the Science Wars, yet not wanting to give up on the valuable findings of the original quest of constructivism: facts are constructed.[4] The path out of the critique's desert, Latour suggests, appears strewn with "nagging doubts" and "tiny cues," one of which is to treat matters of fact as matters of concern. It is in this soul-searching intervention, somewhat gripping in spite of its ironic character, that the notion of MoC appeared as responding to a worry about the disempowering effects—ethical and political—of a constructivism that has too much conceded to the temptations of critique. If science and technology scholars are not outside this intellectual culture, they have, for Latour, good chances to "walk out" from the critical desert. First, because of their technical ways of working: the description of processes and networks on the ground, like "ants" rather than eagles, eschewing a big causal overview. Second, thanks to the nature of its particular subject matter, science and technology, the sturdy "black boxes of science" that even trained as "good critics" the ants couldn't "crack open" (Latour 2004b, 242). Ironically, the invitation to treat matters of fact as *fragile* entanglements of concerns in the so-called human and social sciences draws upon the technical expertise of scholars confronted with the ontological *robustness* of techno-scientific assemblages implying nonhuman agencies.

As I introduced earlier, I read MoC as representing a vision of knowledge politics that Latour attributes to social studies of science and technology. First, an aesthetic: a way of describing things that doesn't split affects, concerns, and worries from the staging of their lively existence. Second, a thingpolitics: a representation of things that gives them a valid voice in the constitution of a "we" by the democratic assembly. Third, a respectful ethos of knowledge production: a critique that doesn't reduce technoscientific things to an effect of (human) power and domination dynamics. Accounting of concerns is a material-semiotic gesture, inseparably a thinkpolitics as much as a thingpolitics. The ethico-political difference that was made by MoC pertains to a knowledge politics that is constitutive of things, not to a dimension of morality that we would add to nonhuman objects and things. However, as I have noted, introducing concern makes a difference across familiar approaches to the politics of things rather than simply confirming them. From an ethico-political and affective perspective, these pertain to an ethos of research and thinking. The assessment of a critical thinking that "runs out of steam," and the proposition of naming facts and things as MoC, responded not only to serious concerns about how things can be misconceived, misrepresented and mistreated but also to the consequences of critical disbelief in science in a worrying world.[5]

It is significant that the notion of MoC was first developed in an intervention addressed to critical thinking in general rather than to critical social studies of science in particular. The appeal was framed within a post-9/11 atmosphere and expressed worry about the contribution of critique to an atmosphere of indiscriminate distrust, in which, even before the "smoke of the event" had settled, "conspiracy theories" rushed to question what *really* happened to New York's twin towers. Such an atmosphere affected of course an intellectual milieu larger than science and technology studies. But Latour's particular call was spoken from the perspective of a collectively learned lesson by this particular academic community. Critical constructivism appeared under the light of lessons learned by researchers and thinkers who had been through the Science Wars—and were accused

of not believing in reality, an idea that Latour himself spent much time trying to counteract (Latour 1999). In other words, the Zeus of critique was
passé, a fairly strawlike figure in a fable's morale rather than an actual problem. Latour's counterparts had moved not only beyond excessively humanist sociopolitical explanations of material and technoscientific worlds but
also beyond suspicious critiques of agonistic interests and power strategies. In sum, Latour was proposing to critical thinkers more generally to
do what STS had, in his view, already learned to do: to respect things as
MoC. According to Latour, STS is at its best in more respectful and, we
could say, constructive ways of exhibiting matters of fact as processes of
entangled concerns. The purpose of exposing how things are assembled,
constructed, is not to debunk and dismantle them, nor is it to undermine
the reality of matters of fact with critical suspicion about the powerful
(human) interests they might reflect and convey. Instead, to exhibit the
concerns that attach and hold together matters of fact is to enrich and
affirm reality by contributing further articulations.

Interlude: Following Power

It can be said that Latour intervened here in a classic question: How are
critical people, in particular researchers, thinkers, and theorists, involved
in the making of the world? MoC emphasizes an ethico-political dimension of that problem: respect for the concerns embodied in the things we
represent implies attention to the effects of our accounts on the life of
things. In other words, if exhibiting the entanglements of concerns at the
heart of human–nonhuman assemblages increases, the affective perception of the worlds and lives we study beyond cartographies of interests and
practical engagements, the staging of a scientific matter of fact or a sociotechnical assemblage, or any other human–nonhuman arrangement as a
MoC, is an ethico-political intervention in its becoming, its mattering.

Yet something troubled my perception of this call to care for the concerns in the things, for things as concerns. An uneasiness with the critique
of critique: In a deeply troubled and strongly stratified world, don't we
still need approaches that reveal power and oppressive relations in the
assembling of concerns? Indeed, beyond exposing a particular thinker's

trajectory to the politics of concern, it appears that this critique of critique is infused with a more general and persistent reluctance—another inheritance of modern science to which scholars, researchers, and academics remain trained in spite of the politization of knowledge—to consider (our) intervention and involvement, and let's say ethico-political commitment and obligations, as an essential part of the politics of knowledge production. So, I had to wonder, could the symmetrical redistribution of affective agency in the complex relationalities of humans and nonhumans in these politics of things reinforce this reluctance?

Thinking with "matters of care" is a way to address this question by offering both contrast and prolongation to the politics of human–nonhuman relations that MoC represent. Matters of care come inspired by feminist thought on care and on knowledge politics. In a way that mirrors my reading MoC, it is also a new notion to speak of old problems by adding problematic layers—reenacting. I couldn't help but read the affirmation of concerns as a delayed effect of preoccupations maintained alive by feminist thinkers—of course not only, but significantly—while others were busy playing critical mini-Zeus. Speculating, with amusement rather than irony, I am tempted to see in increased awareness of concerns a belated engagement with problems enunciated in Haraway's well-known intervention on "Situated Knowledges" (Haraway 1991d), in which she articulated concerns of feminists engaged in a critique of knowledge and science, a contribution acknowledged by general accounts of this field's history (Hess 1997; Sismondo 2010) and continued along frequent reengagements in rethinking its meanings (Mayberry, Subramaniam, and Weasel 2001; Bauchspies and Puig de la Bellacasa 2009). In a piece that strongly influenced inheritor generations of feminist scholars in STS and beyond, Haraway also warned against relying too much on totalizing explanatory theories as well as the corrosive cynicism resulting from mixing deconstructive critique with social constructivism. Among these she included those quests that set out to unmask the truth about how "scientific knowledge is actually made" with particularly power-oriented explanations of scientific and technological success. By these she also referred to early actor-network theory's emphasis on power, but in the form of interest preservation, competition, and agonistic politics, as struggles to settle "the" matter of fact. While this work might have produced good accounts of "how scientific knowledge is actually made," lowly ANTs committed to following *so well* the technoscientific networks ended up telling their story with the same stories and thus reproducing an ethos of agonistic interest battles around knowing (Haraway 1991d, 184; see also Haraway 1994b). In her description of a knowing from nowhere, the trick of a god's-eye view, we can also recognize a Zeus of critical distance, uninvolved and untouched by the wars it causes—or describes. What this implies, however, is that adopting a "following the networks" method—the grounded ant view instead of the eagle one—did not escape the predicaments of being involved in a politics of knowledge. Here, "situated knowledge" did not merely mean knowledge-is-social but also that "our" knowledge is intrinsically politically and ethically situated by its purposes and positionalities—that is, standpoints (Harding 1991). Ignoring this, as feminist scholars had painfully realized, was a way, to translate it in Latour's vocabulary, of lessening reality by erasing or appropriating alternative agencies "from below" (Harding 2008). In other words, taking concerns into account—the ones we study and the ones we have—does go in the sense of a better situatedness. But from a feminist perspective, we cannot teach students that in the task of staging the networks "all this opposition between standpoints and the view from nowhere you can safely forget" (Latour 2005b, 145).

Maybe acknowledging these added dimensions to the politics of taking care of things is why Latour actually called upon Donna Haraway to confirm that MoC need "protection" and "care" (Latour 2004b, 232). With this conception of knowledge politics in mind, the attention to concerns could indeed be seen to modify the affective tonality of the staging of things, opening ways for caring thinking and thinking on caring. So in spite of being born from unease about this critique of critique, matters of care also prolongs MoC. Attention to concerns brings us closer to putting forward the need of a practice of care as something we can *do* as thinkers and knowledge creators, fostering also more awareness about what we care for and about how this contributes to mattering the world. And, as I try to show in this book, we can do so at the heart of distributed agencies in more than human worlds, remaining responsive to material obligations while eschewing moralism and reductive humanist explanations.

Adding Care to Our Concerns

An approach claiming to avoid a moralistic approach to caring knowledge politics asks for some subtlety. A close commentary on how the notion of concern relates to care is an attempt to offer a tactful way to start asking what care can mean for the thinking of things, that is, for the "disobjectified" objects of science and technology. If staging things and matters of fact as MoC adds affective modalities of relation to their reality, how does care in turn affect MoC?

Concern and care have acquainted meanings—both come from the Latin *cura*, "cure." But they also express different qualities. Because of that, however great the importance of care, it does not replace concern at the heart of the politics of things; it brings something else. I have stressed the capacity of the word "concern" to move the notion of "interest" toward more affectively charged connotations, notably those of trouble, worry, and care. As affective states, concern and care are related. But care has stronger affective and ethical connotations. We can think on the difference between affirming "I am concerned" and "I care." The first denotes worry and thoughtfulness about an issue as well as, though not necessarily, the fact of belonging to the collective of those concerned, "affected" by it; the second adds a strong sense of attachment and commitment to something. Moreover, the quality of "care" is to be more easily turned into a verb: *to care*. One can make oneself concerned, but "to care" contains a notion of *doing* that concern lacks. This is because understanding caring as something we do materializes it as an ethically and politically charged practice, and one that has been at the forefront of feminist concern with devalued agencies and exclusions. In this vision, to care joins together an affective state, a material vital doing, and an ethico-political obligation (the entangled aspect of these dimensions and the consequences for ethics are developed further in chapter 4).

As a material concrete doing, care, in order to work, to be deemed adequate or good, is always specific. As Anne Marie Mol puts it, "in the logic of care, defining 'good,' 'worse,' and 'better' does not precede practice but forms part of it" (Mol 2008, 75) and can be recognized in buildings, habits, and machines. This is why the meaning of "caring" can go in different directions, marked by a relation to a range of material practices of historical concreteness. Even among those who agree that "to care" is vital in the worlds of naturecultures and technoscience and who want to bring it to our concern in the representation of things, caring does not necessarily have the same connotations. But the notion of care is also marked by gender and race politics; it brings to mind particular labors associated with feminized work and its ethical complexities. Because of these charged meanings, if "matters of concern" can function as a generic notion for the politics of things (i.e., everything can be potentially thought as a matter of concern), "matters of care" might not. This is not to say that feminist thought should claim a particular ownership around the notion of care but that care is not a neutral notion, nor is a feminist reading of it.

Nuances around the nonneutrality of care can be approached by discussing Latour's particular invitation to care in technoscientific universes. It is in a funny-though-serious dialogue that stages himself speaking to a concerned environmentalist angry with sport utility vehicle (SUV) drivers, where Latour affirms that we need to care for our technologies, even those that we see as pernicious, as *Frankensteinian*—SUVs in his example (Latour 2005c; see also 2007a). Latour argues that it is not a technology that is unethical if it fails or becomes a monster but rather to stop caring about it, to abandon it as Dr. Frankenstein abandoned his creation. Here we can recall Latour's inspiring "scientifiction" on *Aramis* (a promising transport system in Paris) where he tells the story of the collective troubles that led to the abandonment of the project (Latour 1996a). This version of caring for technology carries well the double significance of care as an everyday labor of maintenance that conveys ethical obligation: we must take care of things in order to remain responsible for their becomings.

Recent work that foregrounds the importance of repair and maintenance of technology infrastructures as practices of care supports this case and has expanded it, making a great difference in how objects, devices, and technological infrastructures and the more or less invisible agencies involved in their continuation (Star 1999; Star and Bowker 2007a) are conceived. This work changes the focus on the "robustness" of sociotechnical assemblages, on solid and successful networks or black boxes, by drawing attention to the constant need for repair and maintenance (Jackson 2014; Jackson and Kang 2014), the stakes of their "vulnerable" status (Denis and

Pontille 2014). Here care is approached as myriad laboring agencies and processes—including biochemical interactions and interventions involved in preserving materials and objects exposed to the passing of time and decay (Dominguez Rubio 2016), without which the deceptively unaffected world of material technologies wouldn't turn round. These approaches contribute to thinking sociotechnical agencies through a notion of care that, as Anne Marie Mol proposed, is "not opposed to, but includes, technology" as well as a notion of technology "that is not transparent and predictable, but has to be handled with care" (Mol 2008, 5). Destabilizing and displacing the view of care as "other" than technology opens more than human relations in technoscience to an investigation on the meanings of care. While an ethico-political sense of care is not necessarily explicit here, there is an ethicality at play in the reaffirmation of agencies previously mostly neglected from descriptions of technology. Nevertheless, holding together the different dimensions of care brings an additional question to care as everyday responsible maintenance: its contribution to as well as possible worlds. Affirming that care is necessary to maintain technologies, even technologies that are not necessarily desirable or even harmful, so that they continue to work well opens to further ethico-political interrogations, such as: What worlds are being maintained and at the expenses of which others?

The ethico-political significance of encouraging care for technology orients focus toward a second correlative argument that Latour opposes to the angry environmentalist: that instead of merely criticizing SUVs, if we really want to make a difference about their use, we also have to engage with the concerns that animate those who are in favor of them. This means that to effectively care for a thing we cannot cut off from the composition of its political ecology those we disagree with but who are nevertheless concerned by the thing and the issues it brings to matter. This vision of care is animated by the purpose of treating things as MoC: to engage properly with the becoming of a thing, we should strive to count and include *all* the concerns attached to it, all those who care for it. If we cut off SUV users by demonizing them, not only do we objectify and reduce this socio-material assemblage, by detaching elements of the SUV thing-gathering (machine, producers, and users), but we also become irresponsible: relegated to represent a threatening object, we help to construct SUVs as destructive monsters instead of looking after their possible transformation. Here, care is mobilized to serve a gathering purpose: to hold together a thing and the publics concerned. This "inclusive" vision has political antecedents in inclusive democratic politics (Papadopoulos 2011). But mostly, this way of advocating for care complements the respect for things as MoC with a particular doing: the practical responsibility to take care of the fragile gathering things constitute.

This restaging of the "SUV issue" is a political fable rather than a thick discussion of the intricacies of the case for and against SUV's uses in cities, and my rendering of engages with it in the same way. I am interested in how this way of presenting care goes further than an understanding of care as responsible maintenance of technology. It exhibits an ecumenical version of the "cosmopolitics" of things and of political ecology. As I said before, things are intrinsically political in their own thingy way. When they were hybrids, in *We have never been modern*, the "*tiers état*" (third state) was the political figuration of a misrepresented collective. In that time, things were dwelling and making worlds in the Middle Kingdom but denied agency, objectified *as serfs*—to be used and/or abused by humans/ society. Amusingly, this narrative called upon a powerful and affectively charged political history, the 1789 French Revolution—the "origin" of political modernity—to precisely overrule the binary thinking of the modern constitution and liberate this other *tiers état*: "they [air pump, laboratory] stop being simple, more or less faithful, intermediaries. They become mediators—that is, actors endowed with the capacity to translate what they transport, to redefine it, redeploy it, and also to betray it. The serfs have become free citizens" (Latour 1993, 81). The later lively and more ordinary characterization of hybrids as things/gatherings/matters of concern accounts for these entangled more than human agencies liberated by constructivists. Crucially, their liberation also had to challenge habits of humanistic thinking: "Humanists see the imposture of treating humans as objects—but what they don't realize is that there is an imposture also to treat objects as objects" (Latour 2005d, 6). To become free citizens they have first to be recognized as agents: by doing precisely this, STS scholars played a political role as representatives or liberators of things.

This political role can be crucial to the democratic collective, especially because objectified matters of fact can be used to close down controversies around unsettled political matters. But, as Latour tells us, if something has become an "object," that is, objectified, it is "simply a gathering that has failed—a fact that has not been assembled according to due process" (2004b, 246). This due process is, we could say, the matter of politics when conceived as a politics of things—*dingpolitik* (Latour 2005a). In a later recapitulation, *dingpolitik* designates the multiple negotiated ways through which things come to matter, to become "issues," and to be counted as such (Latour 2007b). Politics appears as a process of progressive inclusions through different stages by which "issues" are processed and assimilated into the cosmos of a democratic society. It works as follows: a new nonhuman entity produces connection and obliges a "cosmogram" to be enlarged or redefined; it generates an issue, a problem, and a concerned and unsettled "public." The machinery of government tries to "turn the problem of the public into a clearly articulated question of common good or common will" and eventually "fails to do so." In this collective machinery, some issues are "metabolized" and "absorbed by the normal tradition of deliberative democracy" and can eventually stop being political, entering the domain of daily routine and administration (817–18). Crucially, this cycle doesn't necessarily close an issue—that is, settle once for all what a thing *is*, ontologically speaking. Some issues hopefully run smoothly—for instance, Paris' sewage system—and don't need to be political. As Marres would put it: no issue, no politics (Marres 2007). But this is not the case for many others, such as SUVs or other controversial technologies (Latour also gives the example of gender politics, which seemed normalized until they were denaturalized, as he notes, by "feminist scholars"). And here Latour calls upon Isabelle Stengers's notion of cosmopolitics (Stengers 2005; Latour 2007b) to designate the process of the different stages of politics of constituting the a-modern, thing-inclusive, democratic assembly. Latour admits, however, that it is the moment of irruption of a challenging entity that pushes to redefine what is the thing, the "issue," at stake that better invokes the cosmopolitical moment: that is, when a gathering is tested about *what it counts as its world* (cosmos). Indeed, for Stengers, this triggers not only processes of inclusion/exclusion but a more *cosmic* concern, a hesitation, a permanent question that challenges the collective by always having as open an unknown: *How many are "we"?*

But one can also recall here that Stengers's cosmopolitical proposal counts significantly among the contestants in the redefinition of issues: those she calls the "idiots"—those who don't want to be "included" or don't even ask to be cared for by a particular assembly, who do not want to become public or cannot "contribute" because they feel that "there is something more important" than the issue at stake, even if that issue might affect their lives too. This may well include "victims" who retain no power to represent themselves as well as radical groups and other unloved others who deeply disrupt, are against, or might fall out of the cycle of representative politics, the cycle of inclusion in an "issue." Without becoming affected by those yet not necessarily *in* the issue, we end up with a purified Cosmopolitics, a leveling of concerns not that far from the misleading Kantian "pacifying" homonym. That this type of detection of indifferent or unresponsive concerns in a gathering is not an easy task, especially if we don't want to merely become "spokespersons" of those who don't ask for it, doesn't preclude representatives of things from trying to learn from such refusals and erasures in order to think carefully through what could be going wrong with the treatment of an issue. In any case, for this ethicopolitical purpose, we would need a displacement of Latour's version of cosmopolitics, which is possibly less concerned than Stengers's by struggles of minoritarian oppositional views.

Obviously the point of Latour's cosmopolitical advocacy for care in this context is not to express particular concern for SUV maintenance and development. Beyond their role as a particular detector of concerns, they just play a part in a political fable on the broader problem of how to do *dingpolitik*. This mode of representing concerns doesn't seem to have a specific stake with respect to the use of SUVs. But then why is the environmentalist opposing this technology staged as self-righteous and close to numb with anger? Maybe the answer to this question resides in the point of departure of this specific advocacy for care, which also puts forward two related problems that Latour has approached elsewhere, too. First, a concern that political ecology in technoscience could remain a marginalized issue, neglected as the problem of a bunch of angry activists instead of a major problem taken into account by contemporary participatory democracies (Latour 2004a)—and therefore an imperative to engage with mainstream politics rather than the "margins." Second, a concern about the pernicious effects on an assembly of those who radically oppose powerful interests sustaining certain technologies, to the point of disengaging with them—here the car manufacturing industry. It is when these oppositions become "fundamentalist" that it becomes more difficult, if not impossible, to give them—SUV haters, for example—a say in an assembly of representative democracy (Latour 2005a). Finally, it could be argued that this way of framing the argument for care in technoscience, like the one to respect concerns, is a response to the agonistic politics of incompatible interests and power relations associated with critical (social) constructivist depictions of technoscience. Read in the wake of Latour's ongoing critique of critique, this type of caring is presented as an obligation of the (environmental) activist to replace excessive critique and debunking suspicions about sociopolitical interests behind things with a balanced articulation of the involved concerns.

Admittedly, if we are thinking from the perspective of these problems, it appears crucial to promote care not only of the technology but also for all those concerned, including those who "care" for SUVs. My problem here is with how the issue is "staged" and, more particularly, how the argument for care is mobilized to protect the "SUV issue" from its objectification by a critical participant—an angry and fairly disrespectful environmentalist. Respect of concerns and the call for care become arguments to moderate a critical standpoint, the kind of standpoint that tends to produce divergences and oppositional knowledges based on attachments to particular visions, and indeed that sometimes presents (its) positions as nonnegotiable— what Latour has sometimes named "fundamentalism." This dialogue thus exhibits mistrust regarding minoritarian and radical ways of politicizing things—here the environmentalist—that tend to focus on exposing relations of power and exclusion rather than just claiming inclusion in the prevailing gathering to reform it from within.

Pausing: Misplaced Concerns

It seems that by asking misplaced questions I have ended up displacing the concerns of MoC. We are now decisively in a terrain of divergence. But why burden a (philosophical) construction with questions that it was not meaning to answer in the first place? To be fair, in terms of knowledge politics the problem that really preoccupies Latour, and for a long time now (Latour 1996a, 19), is somehow wider, methodological: the too-eager "addition" of ready-made "causal" explanations—power structures being one—to "local," on-the-ground, descriptions of a network (the ant's view).

However, this critique comes to be used to debunk arguments that convey minoritarian critical standpoints, elsewhere dismissed as a "eulogy of margins," obsessed with the power of "the center" or, worse, associated with calls for saving "being" from technology (Latour 1993, 122–24). These kind of judgments contribute to form a reductive vision of critical constructivism by ejecting a whole set of concerns from the politics of things. But aren't these critical issues also needed at the heart of an inquiry on science and technology? And couldn't these relate to a non-Zeus-like form of critical constructivism, one that precisely would welcome increased awareness regarding excluded ethico-political and affective concerns? In any case, these are voices required to support a feminist vision of care that can represent concerns with persistent forms of exclusion, power, and domination to which sciences and technologies also contribute. In sum, to promote care in our world we cannot throw away critical standpoints with the bath of corrosive critique.

Other accounts of the life and labors of mediating things can be made possible by those who *make themselves concerned* with marginalized experiences, with the silent annihilation of "unloved others," as Deborah Bird Rose and Thom Van Dooren put it to speak of the neglected lives undergoing silent extinctions (Bird Rose and Van Dooren 2011). Can we not give these a role in the restaging of the mediations that hold things together—even when they might not be easily *detectable* (on the ground) because they have been forgotten, or erased? For those who feel concerned with such issues, and who dedicate efforts to making others care, awareness and civil respect of the voices of concern might not suffice; or we might need to add a layer of care to the concerns of MoC.

Ironically, some of the problems that MoC sets off to address are not alien to feminist issues within the strongly gendered modern bifurcations of nature, segregational splits of gendered beings that early feminist studies of science put so much work into detecting (Bleier 1984; Fausto Sterling

1992; 2000; Keller 1985). A sense of familiarity, even of solidarity, could emerge among the hybrid things of the Middle Kingdom by convoking the memories and struggles of objectified naturecultural feminized entities whose bodies are split by binaries, an argument that Haraway also advances (1994b), as well as Nina Lykke (1996). Yet to account for these concerns about objectification of oppressed "others," we need to care about the deadly monotonous dynamics of power and domination and expand the meaning of care in a politics of knowing. Feminist stories about class, post and decolonial, queer, disability, and antiracist struggles around *what is given into experience* in mediation realms also have plenty to tell about the effects of modern purification (Harding 2008). This is not so much about an obsession to reduce reality to power and domination; it is also about adding layers to reality, in Latour's terms, more reality by further articulation. And this often involves disputing how stories are told in absence of these constituencies. For instance, a feminist, anticlassist, and antiracist account of the liberation of objects-serfs becoming things again would note that, before the 1789 French Revolution, "serfs" (like the marginal and domestics) were actually not even counted as belonging to the *tiers état*—Latour's early political figuration of the Middle Kingdom to liberate. Only counted as *tiers état* were the men who owned land or "goods." These were indeed deemed inferior to the noble class or clergymen, and still underrepresented before the revolution, but not as much as proletarians, women, and people of color (doing the work for those who owned). Excluded from the *tiers état*, they remained barred from democratic representation until well after this revolution.

As Haraway says, it matters what stories tell stories. A staging of the liberation of objects as *tiers état* reproduces the plot: it mobilizes a fleshy and politically charged collective such as the objectified serfs in order to make the *ding* revolution, while forgetting that the actual serfs never got "free citizenship," and how this precious status continues to be a tool for exclusion. These memories of extended networks of domination cannot be dismissed as a human-centered obsession of ready-made political spheres. From a perspective that takes such problems into account, the problematic inheritance of modern humanism is not only an exclusion of objects made *serfs* but an exclusionary arrangement in which the qualifier "human" serves as a measure of objectification, naturalization, animalization, or whatever we call the scandalous things any we—"we" humans or "free citizens"—can do to those it constitutes as "others." There are poisons we cannot just do away with as if they had "never been" (Haraway 1994b)— nor can we wash them away with the bath of humanist politics. Forgetting these stories is to reproduce humanist politics (Papadopoulos 2017). A posthumanist rehabilitation of *things* needs to remember the wider constituency that this word refers to, to think this gathering from the perspective of processes such as what postcolonial thinkers such as Aimé Césaire (2000) call "thingification" and Achille Mbembé, "the body-thing" (2001, 27). What would it mean for a *dingpolitik* to write history with these other mediation doers, these other agents of hybridity? What about the oppressed, suffering, and unhappy "thingified" *servant* beings? It might well be that those don't want to be things, nor issues to others, however respectable the place this name gives to their agency in the democratized networks. Working for a change of perception that makes objects become "things" again might require not only a claim—objects are things again!—but a *reclaiming*. A reclaiming in the sense introduced earlier that doesn't expurgate the stories of minoritarian struggles against thingification but thinks with them in order to problematize the oppressive dynamics involved in bringing a being to qualify as a matter of concern and therefore to deserve (research) attention and care.

Again, one can argue that this way of thickening the staging of thing-realities is simply not the matter of concern of a critique of humanism whose principal concern is including nonhumans in the parliamentary agenda of agenda-setting civil politics. Then why revisit this origin story of the liberation of things? Maybe just to remind us how a seemingly openly inclusive question such as "How many are we?" leaves intact the problem of how to count with agencies that do not fit or cannot even be heard, without transforming politics. I modestly propose a thinking with care that contributes to this effort, to what Dimitris Papadopoulos calls an "insurgent posthumanism" (Papadopoulos 2010) that not only includes new actors but profoundly disrupts and strives to change the conditions of what counts as political agency.

To begin with, working with a feminist notion of care would add layers of concern to the staging of the "SUV issue"—concerns that are not necessarily incompatible to MoC's moderation in the politics of things, but that represent and promote additional attachments as well as create divergence. Care sounds charged to the feminist-attuned not only because of the material practices it signifies but also because they tend to ask critical questions such as who will do something, how and for whom? as well as if, why and how something has come to be devalued. Care convokes trouble and worry for those who can be harmed by an assemblage but might be unable to voice their concern and need for care—for example, trees and flowers, babies in prams whose noses stroll at the level of SUV's exhaust pipes, or whose voice is less heard—cyclists, older people. These ears would hear and even offer sympathy to the anger and frustration of environmentalists trying to bring these experiences to count against successful and robust networks. An account situated by this sense of caring could note that it is not all of "us" that have created SUVs and therefore that there is not a neutral "we" to be held responsible for abandoning this technology to monstrosity. Finally, it would somehow include in the staging of the issue the researcher's own concerns and cares about SUVs and their broader ecological impact: What are *we* encouraging caring for? In other words, if a researcher feels concerned by SUVs requiring care, she/he could stage them in a way that makes others care for their existence: this is the contribution of our knowledge to the production of an oppositional standpoint (Harding 2004). In sum, this account would intervene in how a matter of fact/concern is perceived, prolonged—made to matter in the sociotechnical and naturecultural ontological continuum to which (our) knowledge contributes and is appropriated in, the mattering of technoscience. Posing similar questions, feminist and other critical engagements with science and technology intervene in expanding the signification of caring. This does not mean that only feminist-oriented research holds these concerns but that it offers important resources to explore how thinking with care can affect the problems summed above: the staging of the life of objectified things, their ethico-political representation, and the disempowering affective effects of corrosive critique.

The Erased Doings of Things

Thinking with matters of care understood speculatively is meant as a hopeful gesture, one that might elicit caring ways of thinking, expose them to be prolonged—a purpose that keeps moving in the space-time of writing this book. And yet the notion emerged for me initially as an attempt to show ways in which bringing care into the picture affects the lively life of things. For this I have been inspired by how care is at work in STS even when it doesn't seem to be the focus. Caring is a long-standing concern of feminist thinking, as are objectified beings and the material-semiotic effects of our knowledge politics. Feminist interest in care has brought to the forefront the specificity of care as a devalued doing, often taken for granted if not rendered invisible. I'm thinking, for example, with Lucy Suchman's perspective on projects to develop "smart" interfaces in software "assistant technology." She shows how the search for "autonomous machine agency" and for the artifact that "speaks for itself" contributes to an erasure of "artifactuality." In general, what disappears is "the human labor" involved "in technological production, implementation [and] maintenance." Suchman's account is notably concerned with designs that reinforce the relegation to the shadows of what is considered "domestic," reenacting traditional binaries on the perception of mediating agencies (life upstairs/life downstairs). She shows how these technologies put the needs of the "service economy" at the forefront, reinforcing the "ideal of the independent, self-motivated, entrepreneurial worker" (Suchman 2007b, 219). Smart assistant interfaces are mostly developed in ways that support this ideal by incarnating a "just visible enough worker," who "gets to know us intimately," in order to better accomplish the "superfluous" work so that we can focus on what really counts: the "busy working life." Such designs reinscribe a world where the frailties of assistants must not be noticed: "The litmus test of a good agent is the agent's capacity to be autonomous, on the one hand, and just what we want, on the other. We want to be surprised by our machine servants, in sum, but not displeased" (217–20).

On the one hand, this staging of the liveliness encapsulated in a sociotechnical assemblage provides a better account of the thing "assistant technology" by showing how it reenacts classic distributions of domesticity. Suchman looks out for mediating agencies that would not easily appear in descriptions dedicated to foreground the success of the technology. On the other hand, without contradiction, this is an account that expresses ethico-political attention, an "aesthetics" of staging matters of fact, a politics of representing things: "Our task is to expand the frame, to metaphorically zoom out to a wider view that at once acknowledges the magic of the effects created while explicating the hidden labours and unruly contingencies that exceed its bounds" (Suchman 2007b, 281). Yet in what way do we account for effects that exceed the explicitly gathered concerns of smart assistants, users, and conceivers? Suchman's work asks additional questions such as "What kind of social relations are assumed to be desirable . . . whose interests are represented, and whose labours are erased" (224). In other words, this is not so much asking "how many are we?" but who/what is not counted or not assembled and why, as well as representing the issue with the support of added layers of concerns.

Representing matters of fact and sociotechnical assemblages as matters of care (MoCa) is an intervention. And so is my reading of how Suchman's account turns a sociotechnical issue into a matter of care. First, paying attention to how technological design can reinforce the binaries that devalue domestic/superfluous work contributes to enriching and expanding the inquiry into care as a "sociological" signifier into the realm of things—in continuity with feminist approaches to "domestic" technologies and their role in perpetuating divisions of labor (Wajcman 2000). This gesture involves demonstrating that the issue of devalued ordinary labors that are crucial in getting us through the day shouldn't be treated just as a social or human-centered dimension. It exposes these mediating agencies as not evident, not naturally "reproductive" mediations, but as generative *doings* that support livable relationalities across technoscientific assemblages expands feminist work that has emphasized how agencies of care are not reserved to a particular practice, occupation, or "expression" and are expendable elsewhere. As Silvia López Gil puts it, care includes material and affective tasks related to communication, the production of sociability, and capacity of affect "without which our lives do not work out" and the complexity of which makes them difficult to value, to reduce to a schedule, or to enclose in fixed tasks that "start here and end there" (López Gil 2007). Care all the way down. Returning to the generic notion of care as "everything that we do to maintain, continue, and repair 'our world' so that we can live in it as well as possible. That world includes our bodies, ourselves, and our environment, all that we seek to interweave in a complex, life sustaining web" (Tronto 1993, 103). In the world as we know it, this involves tasks that make living better in interdependence but are often considered petty and unimportant, unproductive, however vital they are for livable relations. And because the doings of care are not restricted to one sociotechnical sphere—for example, to areas such as health care or the responsible maintenance of technology—they require us to look out, as in Suchman's account above, for what exceeds the frame. Potentially, matters of care can be found in every context: exhibiting them appears even more necessary when caring seems to be out of place, or not there—in technical design plans.

Correlatively, Suchman shows how assistant technologies ratify certain everyday tasks as superfluous while they reinforce the superior valuation of "autonomous agency." She refers to how particular forms of design recall the image of slavery: a skillful (self) erasure, "it" must do the mediations but not let us know how vital that work is, how much we depend on it. This recalls the special skills of intimacy to the needs of the owner that Patricia Hill Collins showed were required from the black domestic caretaker, slave, or descendant of slaves (Collins 1986). Feminist insistence on the pervasiveness of care makes a crucial ethico-political and affective matter patent: caring constitutes an indispensable living ground for the everyday "sustainability of life" (Carrasco 2001) and for the survival and "flourishing" of everything on this planet (Cuomo 1997). It is vital: we all need this work, but we predominantly continue to value more the capacity to be self-sufficient, autonomous, and independent from others (Kittay and Feder 2002; López Gil 2007). This is obviously not an uncritical account of technology. It makes a difference for a discussion about the persistence of modes of serfdom in technoscientific worlds. Suchman transforms a sociotechnical assemblage into a matter of care because she *fosters* interest and concern about how a particular human-machine association might further engage humans in neglecting relations of care.

This work shows also how turning a thing into a matter of care doesn't need to be about technology dominating humans nor about ready-made explanations blaming oppressive powers but rather about how a sociotechnical assemblage can reinforce asymmetrical relations that further degrade care, or promote forms of neglect, as they reassemble human–nonhuman agencies. I find another inspiring example in Kalindi Vora's analysis of technologies of transnational IT that allow to offshore affective labor to call centers based in India where people work overnight dedicated to the "customer care" of busy North Atlantic people going about their day (Vora 2009a).

Her critical approach to these redistributions of labor through new technologies expands the discussion on care to a technoscientific assemblage, and again, it does not so much denote an obsession with power and domination but rather a concern about the multiple sites of powerlessness (and strategies of resistance or coping) that trap more or less unseen others at the other end of the telecommunications lines. Making care invisible or externalized doesn't make it disappear. The unequal circulation of care is well documented by feminist research involved in investigating contemporary waged care work, mostly assured by migrant women without legal visible citizenship (Álvarez Veinguer 2008) joining an already dismissed category of unloved workers, whether paid (Duffy 2011; Duffy, Armenia, and Stacey 2015) or not. These, in turn, as shown by the Spanish feminist collective Precarias a la Deriva (Precarias a la Deriva 2006; see also Bishop 2010) will have to develop underground, often "illegal," networks of care in order to overcome the draining of their own living conditions by the burden of care they carry for others with very little recognition.

There is a connection between contributions that expand the meanings of care beyond the sociological and the critical edge this expansion often conveys. One could still ask why should research on care be considered political. In the world as we know it, paying attention to care as a necessary doing still directs attention to neglected things and devalued doings that are accomplished in every context by the most marginalized—not necessarily women—and to logics of domination that are reproduced or intensified in the name of care. Caring, from this perspective, is a doing that most often involves asymmetry: someone is paid for doing the care that others can pay off to forget how much they need it; someone is in measure of caring for somebody who needs care. To represent things as MoCa is an aesthetic and political move in the way of representing things that problematizes the neglect of caring relationalities in an assemblage. Here the meaning of care for knowledge producers might involve a modest attempt for sharing the burden of stratified worlds.[6] This commitment is the political significance of representing MoCa.

Speculative Commitments

Obviously attaching a notion of "matters of care" to such a sociopolitical vision is a thinkpolitics—an intervention. While representing a sociotechnical assemblage as a matter of care can indeed provide a better account of a thing, it also gives ethico-political significance to particular sociomaterial practices by generating care for undervalued and neglected issues. Indeed, other concerns could make such questions irrelevant: Why should we care about these particular erasures? What is wrong with leaving tedious tasks of domestic care to an "assistant" technology so that we can give attention to important things? This is not the point. Nothing might be "wrong"—of course other worlds are also possible and indeed dominant. But invoking care inevitably calls upon constituencies who are or make themselves concerned when technology reinstates interdependency as expendable, or when promising labor-saving devices displace human labor to some unseen elsewhere or by a world in which most laboring "others" have not been replaced by smart digital machines, where their assemblage with these other things intensifies objectification.

Representing matters of fact and sociotechnical assemblages as MoCa is therefore to intervene in the *articulation* of ethically and politically demanding issues. This adds a third dimension to the politics that MoCa tries to convey: not only exposing or revealing "invisible" labors of care in a critical way but generating care. As I have shown when discussing Suchman's work, in strongly stratified technoscientific worlds "erased" concerns do not just become visible by following the articulated and assembled concerns and participants composing a thing. Generating caring might mean counting in participants and issues that have not managed or are not likely to succeed, or even do not want to voice their concerns, or whose voices are less or not perceptible—as agencies of a politics that remains "imperceptible" (Papadopoulos, Stephenson, and Tsianos 2008b; see also chapter 3 of this book). We can think of how legitimate processes of recognition of voiced concerns are consistent within prevalent ways of political understanding and democratic representation focused on facilitating the proper "articulation" of "speech." As Iris Marion Young argues in her analysis of processes of inclusion in democracy, expressing concerns or claims with "articulateness"—often implicitly meaning communicating with dispassionate, formal, general speech—is a dominant form of embodied recognition and participation (Marion Young 2000, 38–39). Alexa Schriempf thickens this argument from the perspective of deaf people and others who have to deal with powerful communication barriers. "Articulateness" means being able to speak in a normalized voice—or to embrace noninnocent technologies that allow access to the world of the articulate (Schriempf 2009). Challenging the predominance of speech in predominant versions of cosmopolitics, Matthew Watson proposes an ethos of listening with care in science and knowledge production as a way to enable "speaking for subaltern epistemic things": "the scientific self emerges as *a mediator listening and giving voice* to an epistemic thing, producing new forms of cosmopolitical cohabitation" (Watson 935–36). Listening, like speaking, is not neutral. Listening with care is an active process of intervening in the count of whom and what is ratified as concerned; it affects the representation of things, adding mediation to mediations. Calls for a more radically democratic way of listening to neglected things speaking "from below" (Harding 2008) could reconnect the politics of care to more than thirty years of discussions in feminist science studies, crystallizing in the argument associated with "standpoint theory" that thinking from marginalized experiences as political (i.e., as problematic) has a potential to transform knowledge (Harding 1991).

Such views bring their own set of difficulties (see chapter 2 for a detailed discussion). One that is particularly important for thinking matters of care in more than human worlds is the framing of this attention as a normative "epistemic" gesture. This view needs some readjustment in an understanding of technoscience where knowers and objects implode, where knowledge is not just "knowledge" but practices and sociomaterial configurations. A materialist conception of care needs to stay close to the implications of caring when giving marginalized things a voice in the staging of technoscientific mediations not only as a way of resisting to idealize care as a moral disposition, but also as a normative epistemic stance disconnected from the material doings that make the web of care in technoscience. When Hilary Rose drew inspiration from women's antimilitaristic struggles and scientific workers' collectives to demand more attention to the concerns and affects expressed by oppositional voices, she engaged with care as it is embedded in fraught material practices for earthly survival. So though Rose framed her gesture as a "feminist epistemology" for the natural sciences, the emphasis on care subverts and rematerializes epistemological questions: "hand, brain and heart" have to work together for alternative sciences and technologies (Rose 1983; see also Rose 1994). Here care represents and foregrounds inseparably neglected and marginalized embodied practices and an ethico-political commitment with oppositional voices involved in the sociomaterial making of technoscience.

Additionally, many discussions regarding the production of ethico-political standpoints have indeed turned around whether these are to be considered as an "epistemological," or even a "methodological," path to include other voices that could make knowledge more accurate (for a collection of these discussions, see Harding 2004). But in the same ways that care as work-affect-politics does not fit well in normative ethics, it also disrupts epistemological norms. First of all, because generating caring standpoints involves much more than creating more accurate knowledge, it is a collective endeavor by which everyday relevance goes beyond scientific validation. Reducing thinking with care to an epistemological stance would constrain its obligations into "a theory" of (good) knowledge and science. But the emergence of caring standpoints is not fostered by normative exhortation. It can be said that standpoints manifest visions that have *become* possible by collective ways of learning to care for some issues more than others—rather than by following a normative ideal. Standpoints come to be through a transformation of habits of perception, thinking, and doing that happen through attachment to particular concerns, interests, and commitments. This is one important reason that situates the knowledge that they influence. In addition, ethico-political standpoints also attempt to add something to the world, something that, we hope, could connect to the gatherings we study in order to make a difference. This involves not only detecting what is there, given in a thing-gathering, but also to think what is not and what *could be*. For all these reasons, standpoints, even when they develop normative tendencies, are not fixed nor essentialist,[7] they depend on material configurations and on our participation in (re)making them. In the same way, an ethos of care in knowledge politics cannot be reduced to the application of a theory of good care; it has to be continuously contested and rethought.

Now, going back to Latour's critique of critical constructivism, we could still ask if this politics of knowledge would be simply an attempt to fit the accounts of things into "ready-made" humanist explanations. We can wonder if a thinkpolitics of care aims simply toward a detection of exploitation, exclusion, and injustice in technoscience. These questions are not unrelated to the pitfalls that Haraway detected in critical (de)constructivism: the consequences of totalizing explanatory visions as well as of indulging in corrosive cynicism about the pervasiveness of power relations. Taking seriously this nuancing of the critical spirit means envisioning a commitment to care for marginalized or neglected issues that is not reducible to suspicious debunking. The ethico-political weariness and disempowerment that self-righteousness of being "on the right side" generates can only aggravate if commitments to oppose forms of power and domination in science and technology are limited to what Latour sees as simplistic (dis) articulations of the world. Convoking concerns that are not present is not simply about adding "ready-made" reasons for their absence—for example, capitalism, gender, race. Fostering care should not become the equivalent of an accusatory moral stance—if only *they* would care!—nor can caring knowledge politics become a moralism disguised in epistemological accuracy: show that you care and your knowledge will be "truer."

Thinking matters of fact as matters of care does not require translation into a fixed explanatory vision or a normative stance (moral or epistemological). I suggest rather that it can be about a speculative commitment to think about how things could be different if they generated care. A commitment because it is indeed attached to situated and positioned visions of what a livable and caring world could be; but one that remains speculative by not letting a situation or a position—or even the acute awareness of pervasive dominations—define in advance what is or could be. Here, too, what care can mean in each situation cannot be resolved by ready-made formulas. It could be said that introducing care in knowledge politics also requires critical standpoints that are *careful*. A beautiful example of this is how Leigh Star and Geoffrey Bowker transformed a suspicious debunking question such as the Hobbesian *cui bono* into a subtle critical detection of the consequences of categorizing some experiences in technoscientific arrangements as "residual" (Star 1995; Bowker and Star 1999; 2007b). Early on, Leigh Star taught us ways of asking *cui bono* that do not set us out on a crusade to uncover conventions and interests sustaining the establishment of exclusions in things. These not only invite us to ask "For whom?" but also "Who cares?" "What for?" "Why do 'we' care?" and mostly *How* to care?" Importantly, these questions can leave open the detection of specific relational arrangements of caring in each situation instead of presupposing there is only one way of caring. As such, they do not totalize—they disrupt totalizations. A commitment to show how forms of domination affect the construction of things and lead to exclusions is not necessarily directed to the disarticulation of the world, or to the negation of the reality of matters of fact and the materiality of technologies, nor even to a reinstating of humanist questions at the *center* of more than human arrangements. Rather, it is a specific way to add to their reality, an urge to getting further involved with their material-semiotic becoming: the coming to matter and ongoing mattering of things.

These knowledge politics are all but a feel-good attitude to caring. On the contrary, they connect caring with awareness of oppression, and with commitments to neglected experiences that create oppositional standpoints. An account of a thing produced with and for care can indeed create divergence and conflict by criticizing the way an issue is assembled. It can produce visions that "cut" differently the shape of a thing, questioning the extension of a network (Barad 2007; Suchman 2007b)—it can even advocate for cutting off components in a matter of concern. Critical sensibility plays a part here, but not in the eagle-like sense of the enlightened unveiling of hidden powers in the real causes of things, not in the sense of a critical distance of a skeptical aspirant to Zeus. A cut does not necessarily generate skepticism and disbelief; it can actually generate more "interest." Not interest in a parochial or agonistic sense but rather in the way thought by Isabelle Stengers (1993, 108): something is interesting if it is situated in-between—*inter-esse*—not to divide, but to relate. This way, the significance of standpoints committed to care is not limited to their critique of power, but also to re-creating relation through that critique. In the perspective proposed here, foregrounding care at the heart of (critical) constructivism, aka ontological politics, reminds us that, for dependent beings, in order to be livable, a critical cut into a thing, the detachment of a part of an assemblage, often involves a reattachment. Not only do we become able to cut in a certain way because of our own attachments—because we care for some things more than others—but also to produce a caring account. Critical cuts don't merely expose or produce conflict, they also foster caring relations. And, in technoscience, thickening Tronto and Fischer's generic definition, this means relations that maintain and repair a world so that humans *and nonhumans* can live in it as well as possible in a complex life-sustaining web.

Affectionate Knowing

Representing things as matters of concern was a response to a bifurcation of nature, a splitting of meanings and matter, the social and the natural in the life of things. In a way, matters of care respond to a related "bifurcation of consciousness" (Smith 1987): the splitting of affective involvements from the researcher's experience. Is there something embarrassing in exposing what we care for? Not only politically embarrassing, but also affectively? It seems that the closer we get to the worlds of science and technology, to the world of "matter," the more we confront something like what Leigh Star called a "Transcendental Wall of Shame." A wall that she found particularly high "when we try to speak of our technological lives in a philosophical manner which includes experience, suffering, or exclusion." We feel it, Star said, when "we are silently shamed—either within academia or within the swamps of convention" (Star 2007). Is care too touchy-feely for the imaginary of technoscientific networks? Is it too suspicious of a naturalization of feeling that seems to contradict the naturecultural entanglements of technoscience? Without forgetting this question, it is helpful to remember that, historically, the "literary technologies" (Shapin and Schaffer 1985; Haraway 1997a) used in accounts of scientific "matters of fact" are meant to sanitize things as matters of fact. These purifications and the silences they produce not only apply to speculative folly, the political, the personal, the petty, and the domestic, but also to embarrassing affections ridiculed in scholarly contexts. Feminist research has often confronted these long-standing habits and their effects in the way science and technology are presented. Making affective engagements an explicit part of the representation of things disrupts these habits of thinking.

This is another dimension of care traditionally neglected in the representation of things: its meaning of affectionate, sometimes loving, connection. A constant source of inspiration for the reintegration of affectivity in how we engage with technoscience and naturecultures has been for me the work of anthropologist of science Natasha Myers. I'm thinking on how she brings into the picture the bodily attachment of molecular biologists to their "objects." Myers shows the crucial affective labor and care involved in "giving life" to molecule models (Myers 2015). What she exposes is that for these "things" to exist, active care and affection are required, not after they are out there as facts, but throughout a process of revealing them as co-generating. With attention to this specific experience of naturecultural relating, she alters the vision that scientists are dispassionately manipulating objects. Prolonging Evelyn Fox Keller's famous phrasing—a feeling for the organism—Myers (2008, 165) notes that "they have a feel for the molecule." These "renderings" of molecular science are all but a detached observation of the human-molecule gathering, far from a cartography of the existing actors and concerns. I have enjoyed seeing Myers presenting aspects of this work in academic contexts. A dancer as well as a scholar, she re-performed the gestures of embodied attachment by which the scientists she studied stage the virtual forms of their molecules. Her writing itself seems to navigate through embodied renderings of care. In a more recent strand of her research, focused on scientists and other practitioners investigating plant sentience, Myers encourages engaging in affective ecologies with plants in which care becomes a basic relational thread, transmitted by the love and passion exhibited by the observer of things, through her own bodily, immersed, involvement in cultivating affectivity. Through involved modes of knowing and writing, Myers not only modifies our perception of the engagements of scientists and their matters, the plants, but she also invites us to become a researcher of a different type: "Here I invite you to cultivate your inner plant. This is not an exercise in anthropomorphism—a rendering of plants on the model of the human. Rather, it is an opportunity to *vegetalize* your already more than human body. In order to awaken the latent plant in you, you will need to get interested and involved in the things that plants care about" (Myers 2013).

I see this compellingly moving invitation as a way of transforming more than human relational arrangements into matters of care, of inevitably becoming affected within them, and transforming their potential to affect others. This meaning of care translated into a way of doing knowledge about science and technology is about finding ways to re-affect an objectified world. Ultimately, as Vinciane Despret (2004, 131) puts it in her beautiful text "The Body We Care For," to "de-passion" knowledge does not give us a more objective world; it just gives us a world "without us," and therefore without "them." Here she refers to the observations of scientists working with animals: the "us" is the human (here a scientist), the "them" the animal. Discussing Konrad Lorenz's experimentation with birds' attachment, she affirms that the passion involved in these relations is not about a "parasitic supplement to some sweet story of love" but about making an "effort to become interested in the multitude of problems presented" to others, interested in what it means "to care." Despret shows also how those who see themselves as carers, and not only as scientists, are affected by those they take care of. She exhibits the ways others care. This is in itself an involved account by which she is encouraging attention to the caring dimensions of knowing. But drawing attention to caring as a form of affectivity in knowledge creation shouldn't be understood as a plea for some form of unmediated love. As Thom Van Dooren (2014) shows, Lorenz's particular modes of doing affectionate care had consequences for birds that became attached to the experimenter who was leading them to "imprint" on him as their primary carer—aka surrogate mother or as mate—consequences that were often dramatic, creating the impossibility of a reattachment to other birds. He notes that it is obvious that Lorenz was caring for his birds, taking roles of parent or mate, feeding them "diligently" and even letting himself feed worms in the process. And yet "all this care cannot be extracted from the broader framework of coercion, captivity, and violence within which it occurred" (105). By interrogating further and prompting to think "that it might not have been so good for geese," Van Dooren engages in a nonidealized vision of caring. However well intentioned toward the things at stake, however interesting the kinds of knowledge it enables, care is a consequential practice that does relationalities as much as undoes them. For what worlds is care being done for? This enlargement of frames does not give a privileged last word to the broader framework, but by engaging with a speculative commitment to care, it is an ethical and political involvement in the ongoing mattering of care.

Looking at these different examples of caring politics in practices of knowing in more than human relations, the question becomes: Can we think of our transformation of matters of fact into matters of care as the doing of carers of a specific kind? Could we, as proposed by Ruth Muller and Martha Kenney (2015), value research methodologies for the caring relational entanglements they produce? Speaking of their interview work with postdoctoral scientists regarding the pressures, constraints, and anxieties they endure, they found that their work "not only produced data but also interfered with the competitive, fast-paced and metric-driven culture of the life sciences in potentially promising ways." Their attitude toward their "subjects" of research became one of solidarity—themselves being postdoctoral students—and an enjoyment in the realization and prolongation of a field of caring relations that overcame disciplinary differences. Here the possibility of care made possible the agency of the "objects" of research as much as that of the researchers. But can we think of these ways of re-creating as well as possible care when the "research subjects" are other than human? And what about when researchers, theorists, and scholars are not in embodied contact with their subjects? Relations of care can take different meanings, but in all of them we also become involved with the matters of fact and the matters of concern. Ways of knowing/caring reaffect objectified worlds, restage things in ways that generate possibility for other ways of relating and living, connect things that were not supposed to be connecting across the bifurcation of consciousness, and ultimately transform the ethico-political and affective perception of things by involvement in the mattering of worlds.

This book has started with an exploration of how an ethico-political concern such as caring could affect the way we observe and present things. I wondered if care in technoscience and naturecultures could mean more than the responsible maintenance of technology and still not become a moral value just added to the thinking of things. These questions pertain to problems of knowledge politics not considered as a separate practice from material worlds in the making. Ways of studying and representing matters of fact and sociotechnical assemblages have world-making effects beyond human existence. The insight that things are matters of concern addresses the ethico-political relevance of constructivist approaches beyond social constructivism and humanist ethics. It also brings us closer to include the importance of care in the life of things, including the affective attachments involved. However, there is a critical edge to care that a politics of gathering concerns tends to neglect. I try to convey this with a notion of matters of care, inspired by feminist contributions to problems akin to those Latour identified in the aesthetic, ethico-political, and affective presentation of the life of things. But matters of care aims to add something to matters of fact/concern with the intention of not only respecting them but of getting further involved in their becoming. It stands for a version of "critical" work that goes further than assembling concerns while aware of the pitfalls of ready-made explanations, power obsessions, and the superimposition of moral or epistemological norms.

Feminist thinking on care both unsettles and enriches the perception of objectified matters of fact. I have gathered work in this first chapter for how it manifests an ethos of care in involvement with scientific and technological assemblages. Caring in this context is both a doing and ethico-political commitment that affects the way we produce knowledge about things. It goes beyond a moral disposition or wishful thinking to transform how we experience and perceive the things we study. Here care stands for necessary yet mostly dismissed labors of everyday maintenance of life, an ethico-political commitment to neglected things, and the affective remaking of relationships with our objects. All these dimensions of caring can integrate the everyday doings of knowledge in and about technoscience.

But the notion of "matters of care" is a proposition to think with: rather than indicating a method to "unveil" what matters of fact are, it suggests that we engage with them so that they generate more caring relationalities. It is thus not so much a notion that explains the construction of things than it addresses how we participate in their possible becomings. Caring here is a speculative affective mode that encourages intervention in what things could be. The constructivist dimension refers to Isabelle Stengers

(2004) in terms of engaging in "constructing a response to a problem." Responding to the weariness of critical constructivism, I wondered if constructivism could contribute, by carefully staging how things hold together, to as well as possible caring relationalities and life conditions in an aching world. But ultimately, what is perceived as a "problem" is always situated, a partial intervention. The initial motivations for this book are provoked by feminist interventions in technoscience that do not see caring as an option but as a vital necessity of all beings, that nothing holds together without relations of care.

This view is embedded in the situational experiences of practice, in the situatedness of the concrete and particular. And yet involving a speculative way to think the ways caring matters in concrete situations of care that ground our interventions could sound counterintuitive. How does the concrete relate to the speculative? What it means is that I'm exploring a generic notion of care without aiming to settle into a coherent concept, into a comforting feeling that worries regarding technoscience would be solved—if only we would *really* care *well* with accurate knowledge of each concrete situation. Because care eschews easy categorization, because a way of caring here could kill over there, we will need to ask "How to care?" in each situation, without necessarily giving to one way of caring a role "model" for others. It means too that as a doing, I look into caring as a transformative ethos rather than a normative ethics. This view remains attuned to ways of knowing on the ground, involved with effects and consequences, with an ethicality involved in sociotechnical assemblages in mundane, ordinary, and pragmatic ways. But formulating the necessity of care as an open question with the potential to transform a terrain from within does add an obligation for the ontological constructivist ethos beyond the power of critique: cultivating a speculative commitment to living worlds. As a transformative ethos, caring is a living technology with vital material implications—for human and nonhuman worlds. The rest of this book is concerned with speculatively enriching this vision of care to engage in transformative thinking and knowing practices in more than human worlds. It aims to incite the question of how to care in ways that challenge situations and open possibilities rather than close or police spaces of thought and practice.

two

Thinking with Care

Reality is an active verb.

—DONNA HARAWAY, *The Companion Species Manifesto*

Corrosive scepticism cannot be midwife to new stories.

—DONNA HARAWAY, *In the Beginning Was the Word*

The epigraphs above disclose that this chapter unfolds as an intimate reading of Donna Haraway's relational ontology, where "beings do not pre-exist their relatings," as a way of exploring how styles of thinking and writing technologies can contribute to relations of care in moving worlds. It is more particularly Haraway's take on the situatedness of knowledge (1991d; 1997a) that I read speculatively as a way of thinking with care. That knowledge is situated means that knowing and thinking are unconceivable without the multitude of relations that make possible the worlds we think with. The premise from which I begin this chapter is thus quite simple: *relations of thinking and knowing require care and affect how we care.* In tune with a nonnormative approach to care as a speculative ethics, the grounds of this premise are ontological rather than moral or epistemological: not only relations involve care, care is relational per se.

Caring and relating share ontological resonance. Again, Tronto's generic definition of care says this well: care includes *"everything that we do* to maintain, continue and repair 'our world' . . . which we seek to *interweave in a complex, life sustaining web"* (Tronto 1993, 103, emphasis added). This vision of caring presupposes heterogeneity as the ontological ground on which everything humans relate with exists: myriad doings—everything we do—and of ontological entities that compose a world—selves, bodies, environment. It speaks of care as a manifold range of *doings* needed to create, hold together, and sustain life and continue its diverseness. This also means that an understanding of human agencies as immersed in worlds made of heterogeneous but interdependent forms and processes of life and matter, to or not to care about/for something/somebody, inevitably does and undoes relation. Its ontological import gives to care the peculiar significance of being a nonnormative necessity. Feminist ethics of care argue that to value care is to recognize the inevitable interdependency essential to the existence of reliant and vulnerable beings (Kittay and Feder 2002; Engster 2005). Interdependency is not a contract, nor a moral ideal—it is a *condition*. Care is therefore concomitant to the continuation of life for many living beings in more than human entanglements—not forced upon them by a moral order, and not necessarily a rewarding obligation.

Of course, not all relations are caring, but very few could subsist without some care. Even when caring is not assured by the people/things that are perceptibly involved in a specific form of relating, in order for them to merely subsist somebody/something has (had) to be taking care somewhere or sometime. Even neglect, the biocidal absence of care, reveals it as inescapable: when care is removed, we can perceive the effects of carelessness. But if care is necessary, it is not given. Speaking of care as (non-moralistic) *obligation* denaturalizes care—for life to even be, it needs to be fostered in some way. That it requires *doing* something indicates not only that it is in its very nature to be about labors of mundane maintenance and repair that require agency (though, as I will argue later in this book, not necessarily intention) but that a more than human world's degree of livability—degree of "as well as possible" living—might well depend on the caring it manages to realize. Standing by the vital necessity of care means standing for sustainable and flourishing relations, not merely survivalist or instrumental ones. Continuing to hold together a triptych vision of care doings-practice/affectivity/ethics-politics helps to resist to ground care as an ethico-affective everyday doing that is vital to engage with the inescapable troubles of interdependent existences.

Haraway's relational ontology has been an inspiration for this journey into care (Puig de la Bellacasa 2004; 2014b) before the theme of care appeared explicitly in her work (2007b, a). First, because for Haraway knowledge and science are relational practices with important material consequences in the shaping of possible worlds. My claim that care matters in knowledge politics—as contributing to the mattering of worlds—is sustained by Haraway's call to pay attention to the workings and consequences of our "semiotic technologies"—that is, to practices and arts of fabricating meaning with signs, words, ideas, descriptions, theories (Haraway 1991d, 187). Following Katie King in recognizing the force of literary apparatuses, Haraway showed us how "bodies" as objects of knowledge are also "material-semiotic generative nodes" (200). Another important source of inspiration are her situated politics of resistance to normativity, both moral and epistemological. These notions are particularly crucial for thinking that intervenes in the more than human worlds of technoscience and naturecultures, with their broken boundaries and imploded worlds where knowledge and ontology collapse. Reading Haraway speculatively is an inspiration for thinking with care in its transformative, noninnocent, disruptive ways.

Thinking-With

Thinking with Haraway is thinking with many people, beings, and things; it is thinking in a populated world. Actually, we could say that for Haraway thinking *is* thinking-with. This singularity can also be read ontologically. "To be many or not to be at all!"—a Harawayan Hamlet could say. Look at the many meanings a word such as "biology" can take in her work: a knot of relationships between living matters and social modes of existence, crafts, practices, and love stories; a range of situated "epistemological, semiotic, technical, political and material" connections (Haraway 2000, 403); an omnipresent discourse; an enterprise of civic education (Haraway 1997a, 118); a metaphor too, but also much "more than a metaphor" (Haraway and Goodeve 2000, 82–83). Objects/bodies of contemporary biology are accounted for as instances of relatedness in the making. This insight goes hand in hand with resistance to reductionism: a constant questioning as to what makes "one." A curiosity about the connected heterogeneities composing an entity, a body, a world, that troubles boundaries: "Why should our bodies end at the skin?" (Haraway 1991a, 178). Haraway's

thinking with thick populated worlds is an acknowledgment of multiplicity but also an effort to actually foster multiplication, to create "*diffraction*":
a politics of generating difference, rather than mere "reflection" of sameness, and the fostering of accountability for the differences we try to make
rather than maintaining a form of moderate "reflexivity" (Haraway and
Goodeve 2000, 101–8; Barad 2007, 71–94).[1]
The way in which Haraway writes is a semiotic technology of these
agitations: connective writing, phrasing worlds together, contributes to
this generative drive. In these incessant web-making moves, ontology is
continuously in the making, in the process of becoming. For Haraway,
"reality is an active verb." This doesn't mean that there are no boundaries
or stabilities but that "beings do not preexist their relatings" (Haraway
2003, 6). Such an affirmation also connects well to work in which, thinking
with Susan Leigh Star, Haraway helped to redefine the "objects" of science
as "boundary projects" (Haraway 1997a, 6).[2] This concerns communities
and collectives, too. For instance, to answer the question of what makes a
feminist "we," we would have to take into account that *feminism does not
preexist its relatings*. Ontologies and identities are affected by collective
politics and positionalities that constantly have to confront and put into
question the boundaries and cuts given in existing worlds (e.g., the taken-
for-granted "woman"). This way of thinking about creating other relations,
other possibilities of existence—namely, other beings—is linked to concerns for the consequences of relations. *What* and *how* we enter in relations affects positions and relational ecologies. No longing here for fixed
realities that could police the outcomes of encounters by confirming correspondence to preexisting "orders."
    A relational way of thinking, which I call here "thinking-with," creates
new patterns out of previous multiplicities, intervening by adding layers of
meaning rather than merely deconstructing or conforming to ready-made
categories. Haraway's work has often reminded me of Gilles Deleuze's and
Félix Guattari's call: . . . it's not enough to shout, "Vive the multiple!" . . . the
multiple *has to be done*.[3] The ways in which Haraway *does* thinking-with-
many has led her to hold multiple ends of sometimes divergent positions,
messing up with preexisting categories. For instance, at the height of hype
surrounding her work, she puzzled attempts to class her as "postmodern"

by affirming: "a lot of my heart lies in old-fashioned science for the people" (Penley, Ross, and Haraway 1990, 9). This resistance to conceptual enclosure is not without political purpose. Giving a fair account of many feminist discussions requires us to cut across fixed theoretical and academic divides. In Haraway's words, "that Hartsock, Harding, Collins, Star, Bhavnani, Tsing, Haraway, Sandoval, hooks, and Butler are not supposed to agree about postmodernism, standpoints, science studies, or feminist theory is neither my problem nor theirs" (Haraway 1997a, 304–5). Refusing the seductions of web disarticulation, and mostly the disengagement with attempting further rearticulation, Haraway identifies the problem "as the needless yet common cost of taxonomizing everyone's positions without regard to the contexts of their development, or of refusing rereading and overlayering in order to make new patterns from previous disputes" (ibid.). There is a cost in dividing and opposing webs of thought that share a history. Readings of conflictive positions in feminist thinking webs in Haraway's writings do not purify the stakes into clear "sides"—this includes fostering efforts to care for each other across conflicts rather than just reinforcing breaks and splits.

But the most striking messing up with categories into which Haraway's thinking-with has drawn her readers is possibly that of inciting us to enlarge our ontological and political sense of kinship and alliance, to dare in exercises of category transgression, of boundary redefinition that put to test the scope of humanist visions of care and thus disrupted existing articulations of concerns. This work welcomes us into a "menagerie of figurations," a "critical-theoretical zoo," where all "inhabitants are not animals" (Haraway and Goodeve 2000, 135–36). Kinships and alliances become transformative connections—merging inherited and constructed relations. This one was never an evident gesture; it carried with it a speculative inquiry that pushed the boundaries of the acceptable. Promiscuous gatherings might provoke unease. So I have seen concerned feminists fairly irritated when Haraway suggested in her influential "A Cyborg Manifesto: Science, Technology, and Socialist-Feminism in the Late Twentieth Century" that we connect with our machines. On the other hand, many postmodern feminists would have rather detached the celebrated cyborg from affections supposed to be essentialist, realist, Second Wave, spiritual, or any other term sounding misplaced in the cyberhype. Look at how the extremely quoted final sentence of her celebrated "A Cyborg Manifesto," "I'd rather be a cyborg than a goddess," has been systematically disconnected from the preceding words affirming that both figures are "bound in the spiral dance"—a characteristic ritual of neopagan activist spirituality for which the figure of the goddess is central (Starhawk 1999). And yet, the jointure of these worlds has crucial significance for speculatively imagining ethical relations that embrace the "other than human" at the heart of more than human ontologies. Daring to connect, reread, and overlayer, Thom Van Dooren vibrantly comments on the compatibility of pagan commitments "to a dynamic, animistic world" with "Haraway's commitment to a dynamic world of active agency in which everything participates, everything acts, in an ongoing process of world making—a process in which all of the various actors literally and physically are the world, as well as being actively involved in the processes and negotiations in which the world takes the specific form that it does. . . . [An] understanding of the world, which acknowledges that nonhuman others—many of whom are often considered to be 'inanimate objects'—are endowed with meaning, power and agency of their own" (Van Dooren 2005). The cyborg is one famous example of attempts to foster relations that might create unease. Purifying these out is a way of policing the possibilities of speculative care. I still remember shocked commentaries at a feminist conference in the late 1990s when Haraway surprised a room's expectant audience by articulating her keynote around stories of personal care for her dog Cayenne. Below I will come back to how attempting to split these unruly webs of caring recalls the bifurcation of consciousness outlined in the previous chapter as the sanitization of affections from scholarly writing and allows neglecting interesting lessons that complicate the affects and responsibilities of caring in ordinary living. For now, what I am attempting to elicit is the style of thinking-with that, in challenging and rejoicing ways, renders such splits difficult to sustain.

Thinking with engagements—of which I have delineated as patterns the politics of solidarities across divergences and the enlargement of the sense of kinship and alliance beyond humanity—goes together with an atypical density in Haraway's writing. Long enumerations exhibit multilayered worlds she both describes and generates. An excess of layering might be a weak spot attached to the singular strength I am associating here to thinking-with. Engaging with inherited worlds by adding layers rather than by analytical disarticulation translates in an effort to "redescribe something so that it becomes thicker than it first seems" (Haraway and Goodeve 2000, 108). "And" is the predominant word of writing-with—before "or," "either," "rather." Situated, implicated, and grounded writing makes it uneasy to skim through, or generalize the claims, especially when writing is deliberately plagued with obstacles to reductionism, to dissection of the webs of relatedness that compose a world. There is no single-issued reading of Haraway because she doesn't write single-issued worlds.[4] This demands from readers an awareness of multiple heritages and an openness to follow lines of surprising connections. It requires an effort to sense how each of her stories is situated in crowded worlds, and it invites a letting go of trying to systematically control a totality. Odd effects occur for readers unfamiliar with the milieus this thinking immerses us in: some amazed and inspired, others can be irritated by a flow of diverse stories and notions and criticize this writing for being obscure.

What this style also invites is a thinking-with committed to a collective of knowledge makers, however loose its boundaries and complex its shapes. A specific meaning of thinking with care appears here that further complicates the reaffecting of knowledge that I approached in the previous chapter: the embeddedness of thought in the worlds one cares for. In Haraway's work this commitment is written-in pretty obviously through a lively politics of quotation that gives credit for many of the ideas, notions, or affects nourishing her thinking: fellow researchers and students, friends, human and nonhuman, beings and forces, affinity/activist groups, whether inside or outside academic or "intellectual" realms. We are often introduced to thick gatherings through a specific event when/where/how an encounter worked for her, changed her, taught her, something. To acknowledge the inscription of a singular thinker in a more than human web is not disregarding idiosyncratic and distinctive contributions to collective intelligences. On the contrary, reading these ways as those of thinking with care is to affirm the worth of a distinctive style of connected thinking and writing that troubles the predictable academic isolation of consecrated authors by the way it gathers and explicitly honors the collective webs one thinks with rather than using others' thinking as a "background" against which to foreground one's own.

The point here does not mean to be hagiographic—indicating remnants in Haraway's work of, for instance, second-wave feminist alternative forms of organizing intellectual work that refused individual authorship—but rather to read this style in a speculative way that can foster the subversive character of thinking with care. Academic institutions do not really value eclectic writing-with, especially when it explodes the category of disciplined "peers" by including unruly affections. Here too a resistance to prefixed collectives is at stake. As Rolland Munro (2005, 250) puts it, what is masked in the "'convention' of publishing whereby academics put their own names to works" is the extent to which it is "the product of a wider collectivity." And authors are not the only instrumental wholes at play in this masking. So are universities. Objectified, separated from each other in order to become "comparable" and enter into competition, academic institutions use complex processes of attribution and reordering to detach the work of their employees from the complex intellectual webs that sustain them—discouraging collaborations within a single department, for instance, in order to be able to underscore the measurable contributions in individual units of work. Only then can thinking and knowledge become individualized *countable* property of an institution. In order to be projected into purportedly manageable futures (e.g., resource allocation), the messy relational entanglements that make our presents need to be "standardized" (Star 1991). What are the consequences of such ordering processes on modes of thought? What delicate threads of the interdependent web of thought and life will be silenced and erased?

The point is not to idealize writing that performs the collective or to suggest that careful quotation will do. Yet seeking ways of inscribing the collective might deserve more attention for its potential to counterbalance the drying effects of isolating academic work. It would be sadly insufficient to reduce these gestures to basic intellectual honesty, academic politeness, or (political) loyalties. What I find compelling in fostering a style of writing-with as a pattern of thinking with care is not so much who or what it aims to include and represent in a text but what it generates: how it actually creates a collective and *populates* a world. Instead of reinforcing the self of a lone thinker's figure, the voice in such a text seems to keep saying: *I am not alone. There are many, many others.* Thinking-with makes the work of thought stronger: it both supports singularity by the situated contingencies it draws upon and fosters contagious potential with its reaching out, its acknowledgment of always more-than-one interdependencies. Writing-with is a practical technology that reveals itself as both descriptive (it inscribes) and speculative (it connects). It builds relation and community, that is: possibility. This way of relating does not speak for creating "unions" or "juxtapositions." These paths follow relation as "something that passes between [the two] which is neither in one nor the other" (Deleuze and Parnet 1987, 10).

This approach also involves resisting a form of academic thinking based on positioning theories and authors in a field by pointing out what "they" are lacking and that "we" come to fill—a puzzle-making approach to critical knowledge. It can alienate those who seek in a text new "data" to complete the (objective) representation of an issue, or conclude on ideas that allow to settle. Moreover, it also troubles the expectation of a "critical insight" that would break with the past by offering a novel pattern emerging out of an obsolete background. But probably the perception most challenged by relationships of knowledge that encourage relations of care might be that affective attachments to collectives are seen as misplaced in academic texts, deemed empathetically uncritical, or even self-indulgent. Skeptical judgments can be particularly acute toward work dedicated to foster commitment to a particular "interpretive community," to what Joan Haran calls "dialogic networks" that "limit the play of reading" and seek common ground for hope in concrete forms of situated "praxis" (Haran 2010). Indeed, much of the trouble with notions of "commitment" is the defiance they inspire in work dedicated to advance specific visions versus a general interest of social description. This is an ongoing challenge for research connected to feminism since the second wave: "politically committed" to a community is identified as "biased." For many feminists, to disrupt this simplification was "fundamental to hopes for democratic and credible science" (Haraway 1997a, 227 n. 3) and a major motivation for the development of feminist "epistemology"—especially of "standpoint theory" as a justification strategy for the knowledge produced from the ground of oppositional movements (Harding 1986). I come back to that particular discussion below, but for now I want to continue articulating the disruptive and creative potential of thinking with care as a way of cutting across existing divides.

Dissenting-Within

Care involved in knowledge making has something of a "labour of love" (Kittay 1999; Kittay and Feder 2002). Love is also involved in compelling us to think with, and for, what we care about. But appealing to love is particularly tricky: idealizations silence not only the nastiness accomplished in love's name but also the work it takes to be maintained. Precisely because of this, it is important to keep in mind that knowledge making oriented by care understood as the labors of love and attachment is not incompatible with conflict, that care is not about the smoothing out of life's asperities, nor should love distract us from the moral orders that justify appropriation in its name (hooks 2000). A nonidealized vision of matters of knowledge creation grounded on committed attachments needs to keep alive the feminist multilayered, noninnocent approach to the loving side of caring.

Relationality is all there is, but this does not mean a world without conflict or dissension. An ontology grounded in relationality and interdependence needs to acknowledge not only, as I said before, essential heterogeneity, but also that "cuts" create heterogeneity. For instance, attached and intense focus on an object of love also creates patterns of identity that reorder relations through excluding some. In other words, where there is relation, there has to be care, but our cares also perform disconnection. We cannot possibly care for everything, not everything can count in a world, not everything is relevant in a world in the same way that there is no life without death. However, I want to suggest that thinking with care compels us to think from the perspective of how cuts foster relationships rather than by how they disconnect worlds. This allows looking at "cuts" from the perspective of how they are re-creating, or being created by, "partial connections" (Strathern 2004). That is, we can draw attention to how "new" patterns inherit from a web of relationalities that contributed to make them possible. The connecting character of thinking that starts from and with the collectives we care for functions as a critical cut—and yet, as affirmed in the previous chapter, one that also speculatively creates "interest" by situating in-between (*inter-esse*), not to divide, but to relate (Stengers 1993, 108).

Affirming that beings do not preexist their relatings means that our relatings have consequences. Multiplying through connection, rather than through distinctive taxonomies, is consistent with a (knowledge) politics not so much driven by deconstruction of the given but to "passionate construction," "passionate connection" (Haraway 1997a, 190). In continuation with attempts for a careful constructivism, engaging in "a better account of the world" is key, rather than limiting ourselves to showing "radical historical contingency and modes of construction for everything" (Haraway 1991d, 187). But if thinking-with belongs to, and creates, community by inscribing thought and knowledge in the worlds one cares for, this is, however, to make a difference rather than to confirm a status quo. By associating thinking-with with relations that make a difference, I am emphasizing prolongations and novel interdependencies more than contrasts and contradictions. Yet for me, thinking with care stems from awareness of the efforts it takes to cultivate relatedness in diverseness, which means, too, collective and accountable knowledge construction that does not negate dissent or the impurity of coalitions. It speaks for ways of taking care of the unavoidably thorny relations that foster rich, collective, interdependent, albeit not seamless, thinking-with. In this spirit I propose an account of two significant interventions drawn from Haraway's work as concrete instances of engagement with the articulations of a caring "we" that conveys learnings from complex conflicts, as vital to thinking-with.

The first account goes back to "A Cyborg Manifesto," as an intervention against "organic" unities. This written manifestation of the unsettledness of feminist history shows how thinking-with can be inspirational, empowering but, mostly, not easy. It contributed to reveal conflicts in feminism as much as proposed alternative narratives of solidarity-building. It stressed how trajectories and positions can connect and transform each other without needing to erase their divergences. A shared urgency manifested in the call: "the need for unity of people trying to resist world-wide intensification of domination has never been more acute" (Haraway 1991a, 154). The proposal was to avoid models of solidarity and resistance to domination that would expect us to rely on evident or given bonding and open ourselves to unexpected "unnatural" alliances: cyborg-coalition politics. The intervention was inspired by and accountable to a wide range of feminist work and activism but more particularly to knowledge and positions conceived within black feminisms and other stances grounded on "oppositional consciousnesses"—in Chela Sandoval's wording (Sandoval 1991; 1995)—that brought radical unease with the ways the multiple situated meanings of "women's experience" were concealed by a white, privileged, and heterosexual feminist "we."

The manifesto, Haraway argued, tried to provoke humor at the heart of something as serious as dreams for political unity (Haraway 1991a, 149). But what I find important to stress here is that this laughter came from an inside, from an involved commitment to problems of a community. This is quite different from the ironical snigger of destructive critique: "I laugh: therefore, I am . . . implicated. I laugh: therefore, I am responsible and accountable" (Haraway 1997a, 182). Laugh with, not laughing at, comes from thinking embedded in communities one cares for, and it is an example of a form of thinking with care that I propose to call *dissenting-within*. And maybe more important, this mode of engaging doesn't only concern visions we are committed to foster. Recognizing insiderness, *withinness*, to the worlds we engage with even if critically is to relate with "complex layers of one's personal and collective historical situatedness in the apparatuses of the production of knowledge" (277 n. 3). This stance is born within complex feminist discussions about the possibilities of objectivity and encourages knowers to not pretend being free of "pollutions" to our vision.[5] And while the example above belongs to *dissenting-within* a collective, testing the edges of a "we," of what we consider "our world," requires also openness to accepting one's thought as inheritor, even of the threads of thought we oppose and worlds we would rather not endorse—as when Haraway describes herself as a daughter of the industrial revolution. Refusing self-erasure of attachments and inheritances is about acknowledging implication, about a way of thinking in interdependency that further problematizes the reverence to critical distance and the correlative value of "healthy" skepticism.

Criticality brings me to a second, and perhaps most salient, example of another "unnatural" alliance in which Haraway's work was intensely involved since the late 1980s and well into the 1990s: the fragile alliance between what Sandra Harding insightfully described as the "women question in science"—addressing the position of women practicing science—and the "science question in feminism"—the feminist critical approach to science as a practice itself (Harding 1991; 1986; Keller 1985; see also Rose 1994). The thorny background for this alliance was described by Londa Schiebinger when reflecting on the split between social studies of science and the sciences they set out to study: "Collaboration became even more difficult when . . . certain factions started practicing intemperate constructivism to the extent that scientists' distrust of scholars examining their disciplines escalated into the 'science wars.'" Schiebinger notes that many feminist researchers developed a refusal of both "reductive constructivism" and "unreflective objectivism." The critical insight that scientific "data," or facts, comport ambiguity due to sociopolitical factors was balanced with respect for the loyalties to "empirical constraints," typical of modern scientific traditions (Schiebinger 2003, 860). Coming back to the weariness of critical constructivism postscience wars discussed in the previous chapter, I am not arguing that feminists have been the only ones involved in seeking more careful forms of critical constructivist approaches to science. But they had also particular reasons for this carefulness that bring forth the difficulties of thinking-with and dissenting-within rather than quarrels around finding out which could be the best normative epistemology to explain the social foundations of scientific practice. Indeed, how could a fertile conversation and alliance take place between radical critiques of science and practicing scientists when critics camp in a position of critical "distance"? How could solidarity with women scientists work out if social scientists claim that their "strangeness" to their field allows them to know better the "native" practitioners' work—that is, enlightening them on the "social" grounds of their so-called natural science (Rose 1996).

With these discussions in mind, a story related to some receptions of Haraway's *Primate Visions* (1992) is particularly touching. Haraway saw this book as an act of love and passionate concern for primatology as a terrain of encounters between multilayered interests. Though it opens with a quote from Eugene Marais: "For thus all things must begin, with an act of love," the book held no illusions of innocence about humans' devouring love for nonhuman others, including the ravages of epistemic love in colonial enterprises set out to research, and hunt, exotic nonhuman and human preys. But neither does the book approach this love with cynicism. Nonetheless, some aspects of Haraway's descriptions made primatologists angry about how their practice was portrayed. Commenting more than ten years later on the adverse reactions to her book by feminist primatologists, Haraway thinks that her ethnographical engagement should have been "thicker," by being more "in the field," and says: "I would have spent more time with my own rhetorical apparatus inviting primatologists into this book—reassuring them. Giving them more evidence that I know and care about the way they think. It became a very hard book for many primatologists. They felt attacked and excluded" (Haraway and Goodeve 2000, 56). This is a quite different claim than ethnographic strangeness as a way toward better, more accurate description. I read these experiences as speaking through Haraway's declared uneasiness not only with social constructivism and deconstructivist approaches to science but also with abstract philosophical realism and critical descriptions from any side disengaged from practicing scientists' concerns (Haraway 1991d). I follow here Hilary Rose, who, in her work about relationships of love, power, and knowledge in feminist science studies, sees the "both/and" positions that Haraway has taken in feminist epistemological debates as a tributary of a "close observation/participation of and in this outstanding group of feminist . . . primatologists" (Rose 1994, 93).

And so I wonder, is thinking from a certain closeness in relations of intervulnerability key for encouraging awareness about the consequences of creating knowledge? To realize that those we set to study and observe are not there only to think-with but also to "live with"? I call upon this wording from Haraway's other manifesto: *The Companion Species Manifesto*. Exploring the "cobbling together" of caring relationality in human-dog love, in the creation of "significant otherness," she affirms: "Dogs, in their historical complexity, matter here. Dogs are not an alibi for other themes. . . . Dogs are not surrogates for theory; *they are not here just to think with. They are here to live with*" (Haraway 2003, 5, emphasis added). This assertion about interspecies love is a sharp acknowledgment of interdependency that embeds thinking with care in relational material consequences. Interspecies love brings additional layers to a concept of more than human modes of care. Care is required in processes in which humans and nonhumans co-train each other to live, work, and play together to construct a relationship of "significant otherness." Haraway's stories about the relations of dogs with humans show that livable relating requires particular care, especially when one of the involved beings depends mostly on the other to survive (Haraway 2007). "Caring for" a nonhuman in a way that doesn't objectify it appears as a particularly noninnocent process involving "non-harmonious agencies and ways of living that are accountable both to their disparate inherited histories and to their barely possible but absolutely necessary joint futures" (Haraway 2003, 7). Care appears as a doing necessary for significant relating at the heart of the asymmetrical relationalities that traverse naturecultures and as an obligation created by "necessary joint futures." Relations of "significant otherness" are more than about accommodating "difference," coexisting, or tolerating. Thinking-with nonhumans should always be a living-with, aware of troubling relations and seeking a significant otherness that transforms those involved in the relation and the worlds we live in.

By speaking of living-with and dissenting-within in the same breath, I want to point to a way of living hand to hand with the effects of one's thinking. Conflicts transform, and continue to transform, the meanings of (feminist) collectives in many places; they challenge our political imagination. Reading moments of dissenting-within as instances of thinking with care stresses the difficulties of taking care of relations involved in knowledge creation. Yet caring for the effects of our thinking—even in worlds we would rather not endorse—can also make us more vulnerable. Recognizing vulnerability as an ethical stance could be an inescapable price of commitment and involvement—if care *moves* relational webs, even by creating critical cuts, those involved in caring are bound to be moved too. Emphasizing the conditions of living-with puts care under the sign of ontological heterogeneity and vulnerability to each other's sort and adds questions such as: How do we build caring relationships while recognizing divergent positions? How do those we study live with the way we think-with them?

What are the effects of knowing practices for other than human significant otherness? Answers to such relational questions are always specific, situated "in vulnerable, on-the-ground work" (Haraway 2003, 7): "There is no way to make a general argument outside the never-finished work of articulating the partial worlds of situated knowledges" (Haraway 1997a, 197). Yet we can still find experiences and stories helpful to learn about the pitfalls of, for instance, well-meaning caring for an "other." And so I end this chapter with a reading of tensions that have presided to feminist visions of positioned knowledge that set out to care for the marginalized, proposing *thinking-for* as an additional feature of thinking with care.

Thinking-For

Reading knowledge through care partially prolongs the argument around which Sandra Harding (1991) gathered the notion of "feminist standpoint theory"—namely, that knowledge committed to "*thinking from*" marginalized experiences could be better knowledge and help to cultivate alternative epistemologies that blur dominant dualisms (Hartsock 1983). This principle has been extensively discussed in relation to feminist reconstructions of women's experiences through oppositional struggle (Harding 2004), but it advocates more generically for a commitment to value knowledge generated through any context of subjugation. A standpoint can be understood as an alternative oppositional vision collectively conceived in the process of tackling situations that marginalize and oppress specific ways of living and knowing. I cannot discuss here the rich and complex genealogies and debates about the meanings and (im)possibilities of this idea.[6] My hope is to contribute to their prolongation by reading them as a form of thinking with care that can be relevant for more than human relations.

It can be said that standpoint as a knowledge politics represents an attempt of people working in academic/intellectual settings to use this space, their daily work, as a site of transformation through the way they produce research and knowledge. Initially thought as an epistemological argument for knowledge producers who belong to communities in struggle (e.g., black feminist women [Collins 1986]), standpoint theory also advocated thinking from marginalized experiences for those who do not necessarily belong to the "margins" in which those experiences are lived; that is, building upon knowledge created in struggles with oppressive conditions. It is on this aspect that I focus here. In Haraway's words: "I believe that learning to think about and yearn toward reproductive freedom from the analytical and imaginative standpoint of 'African American women in poverty'—a ferociously lived discursive category to which I don't have 'personal' access—illuminates the general conditions of such freedom" (Haraway 1997a, 199). This is about knowledge that "casts its lot with projects and needs of those who would not or could not inhabit the subject positions of the 'laboratories,' of the credible, civil man of science" (Haraway and Goodeve 2000, 160). This commitment attempts to connect worlds that do not easily connect, making knowledge *interesting* in the sense emphasized earlier of creating a relation in-between.

And yet this specific way of "thinking from" might be better called a form of *thinking-for* in order to recognize its specific pitfalls such as: confusing ourselves with spokespersons, using marginalized "others" as arguments, or falling into fascination with the inspiring experiences of "the marginal" or the oppressed. This rephrasing also emphasizes the work this kind of solidarity entails—like Tronto's "caring for," it distinguishes the practical concrete commitment to do something from the more self-contained, albeit also consuming, effort of "caring about." Again, the heart of the doing is in the *how* we care rather than the intention or disposition to be caring. Too much caring can be consuming. Women especially know how much care can devour their lives, how it can asphyxiate other possible skills. And care can also smother the subtleties of attention to the different needs of an "other" required for careful relationality. It can be said then that it can also consume the cared for, leading to appropriating the recipients of "our" care instead of relating to them. This translates in yet another reason why creating new patterns by thinking-with requires particular care with our semiotic technologies. Thinking and knowing, like naming, have "the power of objectifying, of totalizing" (Haraway 1991b, 79). In other words, thinking driven by love and care should be especially aware of dangers of appropriation. In fact, the risk of appropriation might be worst for committed thinking, because here naming the "other" cannot be made from a "comforting fiction of critical distance" (Haraway 1991a, 244 n. 4). Groups in struggle can refuse (academic) "speaking for" them as usurpation. Appropriating the experience of an "other" precludes us from creating significant otherness, that is, from affirming those with whom we build a relation. Finally, if thinking with care requires acknowledging vulnerability, this implies that, as approached before in the case of the angry primatologists, our "subject matter," our recipients of care, can answer back. How to care will require a different approach in different situations of thinking-for. Some oppressed "others" do need caring witnesses to act as their spokespersons—for instance, tortured animals in a human dominated world. Caring for the "oppressed" is not an evident commitment. Haraway's hesitations about standpoint theory pointed in this direction: "how to see from below is a problem requiring at least as much skill with bodies and language, with the mediations of vision, as the 'highest' techno-scientific visualisations" (Haraway 1991d, 191).

In her dissenting-within prolongation of Nancy Hartsock's and Sandra Harding's work, Haraway affirmed that a standpoint is not an "empiricist appeal to or by 'the oppressed,' but a cognitive, psychological and political tool for more adequate knowledge." Standpoint here refers to a vision that is the "always fraught but necessary fruit of the practice of oppositional and differential consciousness" (Haraway 1997a, 199; Bracke and Puig de la Bellacasa 2007). Insisting on practice brings us back to the hands-on side of care in the purpose of thinking with others. That is, looking at care as a practical everyday commitment, as something we *do*, affects the meaning of thinking-for. As a privileged woman involved in conversations on the nature of knowledge in feminist science and technology studies I can sincerely acknowledge how much the work is nourished by the risks taken by women scientists to speak out and simultaneously fail to join them in questions such as: How do we actually open the space of science? How do I act in solidarity within the unequal power relationships that keep women from underrepresented groups apart from places I am authorized to work in? We can try to think from, think for, and even think with, but living-with requires more than that. To attempt to multiply the ways of "access," not just to think-for the perpetuated absent. To not confuse care with sole empathy, or with becoming the spokespersons of those discarded. Creating situated knowledge might therefore sometimes mean that thinking *from* and *for* particular struggles require from *us* to work for change *from where we are*, rather than drawing upon others' situations for building a theory, and continue our conversations.

A crucial contribution of standpoint theory to a noninnocent version of thinking with care is that it showed that dismissing the work of care contributes to building disengaged versions of reality that mask the "mediations" that sustain and connect our worlds, our doings, our knowings. It is worth reconsidering an aspect their subsequent framing into an epistemological discussion often obscures: from early on, the "marginalized experiences" on which these theories grounded their visions of mediations were strongly connected to the sphere of care. Dorothy Smith described the everyday domestic material details that a sociologist can ignore in order to be able to write the social *out there*—while sitting in a university office in which the bin has been emptied and the floor cleaned by the invisible night worker. A split that grounds what she calls a bifurcation of consciousness (Smith 1987). Hilary Rose shed light on the work of the invisible "small hands" in laboratories, mostly female, that actually do the sciences as well as called to bring back the heart into our accounts of how science works—the forgotten world of loving and caring absent from most Marxist analyses of work (Rose 1983; 1994); Patricia Hill Collins recalled the black woman's work that provided care to children of slave owners (Collins 1986). Insights to which we could add today's descriptions of the invisible work of migrants often separated from the families they support while they clean the houses and take care of the children of those struggling with better-paid jobs or sweating in fitness clubs to keep up with the exigencies of self-care: all figures of a globalized "chain of care" (Precarias a la Deriva 2004). "Care" here referred us to those layers of labor that *get us through the day*, a material space in which many are trapped. By reclaiming these as a source of knowledge, standpoint theorists were rejecting the epistemic cleansing that obliterates these mediations: a willpower for transcendence that erases everyday actual relations in order to sanitize the production of knowledge, something that Nancy Hartsock (1983) called the production of "abstract masculinity." Thinking of mediating laboring bodies as political (i.e., as problematic) is what standpoint feminisms theorized as a production of positions for building other possible modes of knowing.

Haraway's stance against "political and ontological dualisms" can be read as a continuation of these conversations. The affirmation of the political potential of valuing the world of sticky mediations as a thinking device is prolonged through a generic refusal of *purity*: "The point is to make a difference in the world, to cast our lot for some ways of life and not others. To do that, one must be in the action, be finite and dirty, not transcendent and clean" (Haraway 1997a, 36). I have spoken above of how webs of thinking-with also enact impure connections. This is the meaning of non-innocent thinking, of "staying with the trouble" (Haraway 2016)—not critique that sets us on the "right" side. The disruptive potential of thinking with care to keep us close to the earthy doings that foster the web of life remains active in these efforts. Care remains a terrain for these reaffirmations. I am thinking of a spirited response in Haraway *When Species Meet*, where she confronts the fascinatory effects of Deleuze and Guattari's "becoming-animal" as they embrace the figure of the wolf—the pack, the multitude of deindividualized *affect*, the gate to the wild—opposing it to dismiss the domesticated dog as the focus of petty, sentimental, familial, and regressive *affection*: epitomized by an old lady's dog. Haraway forcefully rejects here "a philosophy of the sublime, not the earthly, not the mud" that manifests "scorn for the homely and the ordinary." What strikes Haraway in the distaste for the "little cat or dog owned by an elderly woman who honours and cherishes it" is the "disdain for the daily, the ordinary, the affectional" (29); a "display of misogyny, fear of aging, incuriosity about animals, and horror of the ordinariness of flesh" (Haraway 2007b, 30). A more than human thinking with care would cherish every insight for alternative relatings to be found in the worlds of domestic, petty ordinariness, the difficult and playful, the joyful and aching mediations of caring affection, crucially involved in everyday experiences of interspecies intimacies in contemporary naturecultural worlds. It would not try to split these from the spheres of awe-inspiring thinking of the posthuman; it would consider continuous ordinariness, as much as the worth of thinking, as the eventful breaks with the conventional.

Haraway's work to hold contradictions and complexities together rather than purifying them is a rich resource for those seeking to prolong feminist engagements in disruptive care. This is continued in her most recent work on interspecies relationalities. With this thinking ethos she explores the predicaments of caring in a naturecultural world by showing how embedding caring in more than human futures could well involve embracing unexpected attachments that may seem revulsive (Haraway 2007b). I was first gripped by the relevance of Haraway's thought to work with the particular impurities of caring in naturecultures when she posed the compelling question "Which is my family in this world?" (Haraway 1997a, 16, 51). The question was required by a particularly unsettling cyborgian creature: a transgenic rodent, Oncomouse™, produced to serve research on breast cancer. Caring for this mouse is an unusual experience, at least the way Haraway retold her story, steered away from any temptation of sentimentalism. Named both a "she" and an "it," her specious boundaries are impure, she lived in labs but she was not a mechanical device, she suffered, but she was not "just" a collateral effect of the experimental setting: she was serially born-produced-patented to suffer. By dying or surviving, Oncomouse™ was supposed to *prove* what type of being is cancer. But by thinking-with Oncomouse™'s life from a feminist perspective, by asking speculatively committed questions such as for whom Oncomouse™ lives and dies, Haraway's testimony, illustrated with Lynn Randoph's effective portrait of a naked martyr mouse wearing a crown of spines and under constant observation in a peep-show lab, also "proved" something unexpected: our sister mouse was made to play a part in the conglomerate of industrial, medical, and economic interests that constitute the "cancer complex" (Jain 2013). Vis-à-vis such beings, and of these kinds of technosciences, the feminist sense of caring was urged to mutate, and maybe more than ever. Oncomouse™ was an edifying story of antisignificant otherness provoking us into a sense of enlarged sorority. Looking at experimental ways of life through the eyes of our abject sister mouse revealed the persistent ethos of the *disinterested* modest witness in the experimental laboratory as the utmost uncaring insult. Upsetting the illusions of modern science by forcing us to look through the eyes of this high-tech lab rat, Haraway disrupted a matter of fact into a matter of care.

This account adds significance to thinking with care in more than human worlds. It brings us back to the joint fortunes of all forms of life with sociotechnological becomings, the grounds on which this book started its trajectories. Haraway's commitment to tell stories in ways that emphasize noninnocent relations contributes to the ongoing reenactment of a politics of care as an everyday practice that refuses moral orders that reduce it to innocent love or the securitization of those in need. Adequate care requires a form of knowledge and curiosity regarding the situated needs of an "other"—human or not—that only becomes possible through being in relations that inevitably transform the entangled beings: living with is for Haraway a becoming-with. If I have insisted so much on the politics of acknowledging loving involvements, it is because I believe Haraway's accounts of these co-transformations have been made stronger and more able to generate matters of care by how she shares her own intimate relationships with, for instance, the dog she mostly cares for; by how she exposed her own transformations through their relationship and embeddedness in a collective layered history of ethical predicaments. By embedding relations of care in situated entanglements, she shows that responsibility for what/whom we care for doesn't necessarily mean being *in charge*, but it does mean being involved.

Pausing: How Are You Doing?

I have started this book by making the case for the meaningfulness of care for thinking and knowing. None of these features—thinking-with, dissenting-within, and thinking-for—are meant to promote a ruling for ethical knowledge. I am not arguing that every account of relations should represent care through this way of writing, only that caring engagements shouldn't be dismissed as accessory. I believe it's important to resist enrolling care for a normative theory of knowledge. If there is an ethics and a politics of knowledge at stake, it cannot be a theory that would serve us as a "recipe" for doing our encounters. I have invoked speculative thought as a way of conjuring normativity, both moral and epistemological. Insofar as we remain committed to ongoing curiosity with the specifics of "how" it could be done, care is a good trope to exhibit the singularity of a nonnormative politics, and ethics, of knowledge. Reminding one of the suggestions this book is trying to explore, to think the different dimensions of care together and through each other, caring affection, as something we *do*, is always specific; it cannot be enacted by a priori moral disposition, nor an epistemic stance, nor a set of applied techniques, nor elicited as abstract affect.

And yet as I attempt to give significance to care in knowledge relations, I feel that the pitfall to look out for remains the traction of epistemological moralism. Something holds together, something matches—something feels *true* enough as for trying to impose it, to convince. Maybe there is no wonder: the term "accurate" derives from care, "prepared with care, exact"; it is the past participle of *accurare*, "take care of." Here, the notion of doing something with care led to that of "being exact." Without being cynical about desires to be true, to be just,[7] the tempting proximity between these terms reveals a risky ground: the ambition to control and judge what/who/how we care for. This controlling aim echoes what happens with purposes of collecting knowledge practices under normative epistemologies that tend to erase the specificities of knowing practices. How do we keep thinking with care from falling in a *too much*, into a devouring will for controlled accurateness, to be *all* right?

Haraway's knowledge politics thicken and complicate the meanings of caring for thinking and knowing precisely because they enact resistance both to epistemological formatting and to tempting "orgies of moralism" (Haraway 1997a, 199) as solutions to sort out once for all the difficulties of significant interdependency. Maybe her antidote to normativity itself, whether epistemological or moral, is an appetite for unexpectedness pervasive in her ontological webbings: "I am more interested in the unexpected than in the always deadly predictable" (280 n. 1). And because, in her words, "nothing comes without its world," we do not encounter single individuals, the meeting produces a world, changes the color of things, it diffracts more than it reflects, distorts the "sacred image of the same" (Haraway 1994a, 70). Knowing is not about prediction and control but about remaining "*attentive* to the unknown knocking at our door" (Deleuze 1989, 193). Encounters have unexpected outcomes: "What is it that happens precisely when we encounter someone we love? Do we encounter somebody, or is it animals that come to inhabit you, ideas that invade you, movements that move you, sounds that traverse you? And can these things be parted?" (Deleuze and Parnet 1992 [1977], 17). We do not always know in advance what world is knocking, or what will be the consequences, and yet *how* to care remains a question on how we relate to the new. Foucault once reminded the etymological acquaintance of care with "curiosity," to revalue the latter as "the care one takes of what exists and what might exist" (quoted in Latimer 2000). Haraway has often called for engaged curiosity as a requisite of better caring for others in interspecies relations (Haraway 2007b). Therefore it is not difficult to see how cyborgs and other hybrid beings can be called to support the importance of care in more than human worlds, not only because these extend meanings of caring out of expected normalized forms of kinship to embrace the unfamiliar—pace Latour, *frankenstenian*—forms of life emerging in technoscience, but more generally because this gesture reveals that thinking with care can never be settled, one theory won't do the job in the worlds that come with Haraway's speculative writing: demands for caring will not cease to come from "unexpected country" (Haraway 2007a). This shows us that the task of care is as unavoidable as always ongoing, new situations change what is required of caring involvements.

Thinking with care as living-with inevitably exposes the limits of scientific and academic settings to create more caring worlds. I pause in my exploration to present a basic curious question to my readers: *How are you doing?* I would like this question to sound like a mundane way of caring, within a respectful distance, for what/whom we encounter and we don't necessarily know, a communication device required for thinking with care in populated worlds. It could indicate curiosity about how other people keep care going in the dislocated world of contemporary academia and other fields of controlled and technoscientific knowledge production with its managerial corollary, the anxious delirium of permanent reorganization: "I can't go on. You must go on."[8] Thus, "How are you doing?" sometimes might mean "How do you *cope*"?

Some would say that producing knowledge that cares is mostly about "caring about," requiring less hands-on commitment than concretely toiling in the worlds that we study, "out there." Yet having proposed to embrace a certain form of vulnerability in knowledge engagements might require also acknowledging that these can take their toll. The affective tensions of care are present in its very etymology, which includes notions of both "anxiety, sorrow and grief" and "serious mental attention." Or one could wonder, aren't anxiety, sorrow, and grief actual threats to the serious mental attention required by thinking with care? Does the attention required to keep knowledge aware of its connectedness and consequences inevitably lead to anxiety? A major pitfall is that too much caring can asphyxiate the carer and the cared for. But can this prevent us from caring? Aren't anxiety, sorrow, and grief unavoidable affects in efforts of paying serious mental attention, of thinking with care, in dislocated worlds? Or do these affects belong to an out-of-place sense of *inaccurateness*; the sense that something does not match, does not hold together; the feeling driving speculative thinking that something could be different?

A politics of care goes against the bifurcation of consciousness that would keep our knowledge untouched by anxiety and inaccurateness. Involved knowledge is about *being touched* rather than observing from a distance. Starting from this premise, the next chapter explores the meanings of knowing as touch, as a *haptic* technology that questions the modern humanist transparency of (distant) vision. It follows contemporary engagements with technologies of touch that are rejecting the primacy of vision in traditional epistemologies, it addresses the desire for thinking in intimacy, in proximity with the mediations that make the world possible. Touch therefore opens further meanings of knowledge that cares. And yet my conversation with reclamations of the haptic sensorial universe becomes itself an attempt to withstand the idealizing of care as a more immediate form of knowledge. While touch is maybe the sense that best embodies the involved intensities of caring doings and obligations, speculative thought on the possibility of care troubles longings of immanent immediacy.

❖❖

three

Touching Visions

The affective, ethical, and practical engagements of caring invoke involved embodied, embedded relations in closeness with concrete conditions. And yet I am exploring care for a speculative ethics. Embracing the tension between the concrete and the speculative, this chapter engages with paths to the reembodiment of thinking and knowing that have been opened by passionate engagements with the meanings of "touch." Standing here as a metonymic way to access the lived and fleshy character of involved care relations, the *haptic* holds promises against the primacy of detached vision, a promise of thinking and knowing that is "in touch" with materiality, touched and touching. Yet the promises of this onto-epistemic turn to touch are not unproblematic. If anything, they increase the intense corporeality of ethical questioning. In navigating the promises of touch, this chapter attempts to exercise and expand the disruptive potentials of caring knowing that this book explores. It attempts to treat haptic technologies as matters of care, and in doing so continues unpacking and co-shaping a notion of care in more than human worlds.

Unfolding and problematizing the possibilities of touch draws me into an exploration of its literal as well as figural meanings. I follow here the enticing ways opened in theory and cultural critique to explore the specificity and interrelation of different sensorial universes (Rodaway 1994; Marks 2002; Sobchack 2004; Paterson 2007). All senses are affected by these reexaminations of subjectivity and experience, but touch features saliently, as a previously *neglected* sensorial universe, as a metaphor of intensified relation. So why is touch so compelling? And what new implications for thinking are being suggested by invoking touch?

Attention to what it means to touch and to be touched deepens awareness of the embodied character of perception, affect, and thinking (Ahmed and Stacey 2001; Sedgwick 2003; Blackman 2008). Understanding contact as touch intensifies a sense of the co-transformative, in the flesh effects of connections between beings. Significantly, in its quasi-automatic evocation of close relationality, touching is also called upon as the experience par excellence where boundaries between self and other are blurred (Marks 2002; Radcliffe 2008; Barad 2012). The emphasis on embodied interaction is also prolonged in science and technology studies, for instance, by exploring "the future of touch" as made possible by developments in "robotic skin" (Castañeda 2001). Drawing attention to laboratory touching devices can also highlight the materiality and corporeality of subject-object "intra-actions" in scientific practices, missed out by epistemologies founded on "representation" that tend to separate the agencies of subjects and objects (Barad 2007). Touch emphasizes the improvisational "haptic" creativity through which experimentation performs scientific knowledge in a play of bodies human and not (Myers and Dumit 2011, 244). And engaging with touch also has political significance. In contrast to expecting *visible* "events" that are accessible to or ratified by the politics of representation, fostering of "haptic" abilities figures as a sensorial strategy for perceiving the less noticeable politics in ordinary transformations of experience missed by "optic" objectivist representation (Papadopoulos, Stephenson, and Tsianos 2008b, 55). Here, haptic engagement conveys an encouragement for knowledge and action to be crafted *in touch* with everyday living and practice, in the proximity of involvement with ordinary material transformation. I read these interventions as manifesting a deepened attention to materiality and embodiment, an invitation to rethink relationality in its corporeal character, as well as a desire for concrete, tangible, engagement with worldly transformation—all features and meanings that pertain to the thinking with care that I am exploring in this book.

Embodiment, relationality, and engagement are all themes that have marked feminist epistemology and knowledge politics. Exploring meanings of touch for knowledge politics and subjectivity prolongs discussions

regarding situated and committed knowledge initiated in chapters 1 and
2. To think with touch has a potential to inspire a sense of connectedness that can further problematize abstractions and disengagements of
(epistemological) distances, the bifurcations between subjects and objects,
knowledge and the world, affects and facts, politics and science. Touch
counteracts the sensorial metaphor of vision, dominant in modern knowledge making and epistemologies. But the desire for better, profounder,
more accurate vision is more than a metaphor. Feminist critiques have
questioned the intentions and the effects of enhanced visual technologies
aimed at penetrating bodies to open up their inner truths.[1] Engaging
within this tradition of ontopolitical suspicion about visual representation, Donna Haraway proposed nonetheless that we reappropriate the
"persistence of vision" as a way to engage with its dominant inheritance.
The challenge is to foster "skill . . . with the *mediations* of vision" (Haraway
1991d, 191, emphasis added), notably by contesting and resisting to adopt
an unmarked and irresponsible "view from nowhere" that pretends to see
everything and everywhere. This embodied and *situated* material-semiotic
reclaiming of the technologies of vision is at the heart of her reworked
figure of a "modest witness" for technoscience (Haraway 1997b) that trans-
figures the meanings of objectivity in ways that opens possibilities for
knowledge practices committed to as well as possible worlds (Haraway
1991d, 183–201).

Significantly, by embracing touch, others have also sought to emphasize
situatedness and make a difference in cultural atmospheres strongly attuned
to visual philosophical models of ways of being in the world (Radcliffe
2008, 34). Is knowledge-as-touch less susceptible to be masked behind
a "nowhere"? We can see without being seen, but can we touch without
being touched? In approaching touch's metaphorical power to emphasize
matters of involvement and committed knowledge, I can't help but hear
a familiar voice saying "theory has only *observed* the world; the point is
to *touch* it"—lazily rephrasing Marx's condemnation of abstract thought
that "philosophers have only interpreted the world . . . the point is to
change it." And yet, the awareness, suggested in previous chapters, that
knowledge-making processes are inseparably world-making and materially consequential does evoke knowledge practices' power to touch—and

commitment to keep in touch with political and ethical questions at stake in scientific and other academic conversations.

Engaging in discussions that are revaluing touch brings me back to the paradoxes of reclaiming. Reclaiming technologies of vision entailed reappropriating a *dominant* sensorial universe and epistemological order, seeking for alternative ways of seeing. The poisons encountered in these grounds are optic arrangements that generate disengaged distances with others and the world, and claims to see everything by being attached nowhere. In contrast, much like care, touch is called upon not as dominant, but as a *neglected* mode of relating with compelling potential to restore a gap that keeps knowledge from embracing a fully embodied subjectivity. So how, then, is reclaiming touch opening to other ways of thinking if it is already somehow an alternative onto-epistemic path? The reclamation of the neglected is in continuation with the thinking strategy encountered in the previous chapters: thinking from, with, and for marginalized existences as a potential for perceiving, fostering, and working for other worlds possible. But these ways of thinking don't need to translate in expectation that contact with the neglected worlds of touch will immediately signify a beneficial renovation. On the contrary, to reclaim touch as a form of caring knowing I keep thinking with the potential of marginalized oppositional visions to trouble dominant, oppressive, indifferent configurations, a transformative desire that also requires resisting to idealization. When partaking in the animated atmosphere of reclamations of touch, there is a risk of romanticizing the paradigmatic other of vision as a signifier of embodied *unmediated* objectivity. Rather than ensuring resolution, thinking with touch opens new questions.

The Lure of Touch

Like others, I have been seduced into the worlds of touch, provoked and compelled by the very word, by the mingling of literal and metaphorical meanings that make of touch a figure of intensified feeling, relating, and knowing. Its attractiveness to the project of this book, however, is not only that of evoking a specifically powerful sensorial experience but also that of providing the affective charge that makes it a good notion to think about the ambivalences of caring. Starting with being touched—to be attained, *moved*—touch exacerbates a sense of concern; it points to an engagement that relinquishes detached distance. Indeed, one insight often advanced about the specificity of experiencing touch (often supported by references to Merleau-Ponty's phenomenology) is its "reversibility": when bodies/things touch, they are also touched. Yet here already I wonder: to touch or to be touched physically doesn't automatically mean *being in touch* with oneself or the other. Can there be a detached touch? Unwanted touch, abusive touch, can induce a rejection of sensation, a self-induced numbness in the touched. So maybe we have to ask what kind of touching is produced when we are unaware of the needs and desires of that what/whom we are reaching for? This resonates with the appropriation of others' through caring that I discussed in the previous chapters; the troubling character of these dynamics is exacerbated when thought can be conceived as a corporeal appropriation through "direct" touch.

These questions become more pressing when facing touch's potentially *totalizing* signification: touch, affirms Jean Louis Chrétien, is "inseparable from life itself" (Chrétien 2004, 85). I touch, therefore I am. There is something excessive in that we touch with our whole bodies, in that touch is there *all the time*—by contrast with vision, which allows distant observation and closing our eyes. Even when we are not intentionally touching something, the absence of physical contact can be felt as a manifestation of touch (Radcliffe 2008, 303). Moreover, to be felt, sensorial and affective inputs that other senses bring to experiencing necessarily pass through material touching of the body. This total influence contributes to a sense of "immersion" (Paterson 2006, 699) and is incarnated in its atypical, all-encompassing organ, the skin (Ahmed and Stacey 2001). Touch exhibits as much ascendancy as it exposes vulnerability.

Touché is a metaphorical substitute for wounded. The way in which touch opens us to hurt, to the (potential) violence of contact, is emphasized by Thomas Dumm, who reminds us that touch comes from the Italian *toccare*, "to strike, to hit." Dumm's meditations on touch are particularly illuminating regarding its ambivalent meanings.[2] Touching, he says, "makes us confront the fact of our mortality, our need for each other, and, as [Judith] Butler puts it, the fact that we are undone by each other" (Dumm 2008, 158). In contrast, Dumm explores two meanings of becoming *untouchable*.

First, the loss of somebody we cared about that makes this person untouchable: "that which we imagine as part of us is separate now" (132). Second, to become oneself untouchable: "a figure of isolation, of absolute loneliness" (155).

But how would becoming untouchable, to undertake a protective disconnection with feeling, be possible given the omnipresence of bodily touch? Total presence of touch doesn't necessarily entail awareness of its influence. Dumm makes us see that rejecting touch is possible and sometimes necessary to survive hurt. Yet if such shielding becomes entire, it entails a negation of life itself. The unavoidable ambivalence of touch is thus of conveying a vital form of relation and a threat of violence and invasion. Dumm unfolds Ralph Waldo Emerson's avowal of feeling untouched by the death of his son and his affirmation that touching is both "an impossible act" and necessary for becoming "actors in the world of experience." Dumm concludes that losing touch is a flight into the "futility of total thought," while touching is a turn to the "partial nature of action," a move "from transcendence to *immanence*, from the untouchable to the embrace of *corporeal* life" (Dumm, 158, emphasis added). Life is inevitable mortality, partiality, and vulnerability: the troubles and *conditions* of living. Trust might be the unavoidable condition that allows this openness to relation and corporeal immanent risk.

Exposure through touch translates into another emblematic extreme often associated with touch, healing: "If I only touch his garment, I shall be made well," thinks a sick woman approaching Jesus (Matthew 9:21). This biblical verse came to mind as I encountered the logo for a company developing three-dimensional anatomical simulation software for medical learning purposes—TolTech—*Touch of life technologies*.[3] It featured two human hands, index fingers extended to touch each other, invoking the divine connection between God and Adam represented by Michelangelo and his apprentices on the ceiling of the Sistine Chapel. However, offering "the ability to approach the human body from any combination of traditional views," Touch of Life's version referred to the enhanced vision of anatomical parts via 3D technologies that could bring medical practitioners in training closer to a re-creation of actually touching them. The image was science-fiction oriented, offering a first-contact extraterrestrial-like sight of two index fingers at the point of touching, contrasted against an outer-space dark blue background. An uncanny light had been depicted emanating from the space close to this not-yet-accomplished contact, producing circling waves of brilliance that contour supernatural hands. The technobiblical imagery invoked by this vision of medical technology appealed to ancestral yearnings of healing transformation, and maybe salvation, through embodied and direct contact with a powerful technoscientific (godlike) promise.

Touch is mystical. Touch is prosaic. Though neither scientific nor political cultures have ever been (totally) secular, there is, however, a sensible way in which embodied contact with *evidential knowledge* is associated with the material rather than the spiritual. This connection is supported by a long history in which concrete, factual, material knowledge is opposed to "bare" belief. Remaining in the biblical imaginary, we can remember Saint Thomas, who became the paradigmatic doubter, manifesting human weakness in his need to touch Jesus in order to believe the news of his resurrection. In declarations following the explosion of the financial speculative bubble leading to the 2008 financial crisis, Benedict XVI, the catholic pope in office at the time, encouraged people to hold on to beliefs that are not based on material things. He warned that those who think that "concrete things we can touch are the surest reality" are deceiving themselves.[4] This time, touch falls decisively on the side of prosaic knowledge; it serves the doubtful, those who need to get hold of something, while faith belongs to trust in untouchable immaterial forces. During the first years of the crisis, my bank was nationalized after it threatened to collapse. It struck me how, months later, its offices still exhibited posters of a campaign inviting clients to give up "paper titles" in favor of digitalized ones with the slogan: *Dematerialisation. Inform yourself here.*[5] Pope Benedict XVI was clearly out of touch with what critics of the imploded financial system had been relentlessly highlighting: the immaterial and unreal character of a speculative bubble frantically inflated by global markets disconnected from the finite material resources of people and this planet. Dematerialized, financialized wealth. From this perspective, it was not so much the materiality of things we can touch that led to the global financial meltdown in 2008 but their deadly negation by a "delirious," out-of-touch capitalist version of the speculative (Cooper 2008).

My point here is not to refute faith in the ungraspable, nor the appeal of touching the concrete. I am just realizing how easily an inclination for touch as a way of intensifying awareness of materiality and immanent engagement can get caught in a quarrel about what counts as real and authentic, worth of belief and reliance. Whether this "real" is a source of divine promise or of tangible factuality, *authenticity* is at play. This aspiration to the truthful is reproduced by promises of enhanced immediacy and intensified *reality* in computing experience that abound in the research markets of innovative haptic or touch technologies. If seeing stands for *believing*, touching stands for *feeling* (Paterson 2006). Here, *to feel* becomes the ultimate substantiation of reality, while seeing is expelled from genuine feeling, and believing's authenticity rate plummets. The rush to the "material" in reclamations of touch made me wonder if the increased desire for touch manifests an urge to rematerialize reliability and trust within a technoscientific culture fueled by institutionalized skepticism? In other words, could the yearning for touch manifest also a desire to reinfuse *substance* in more than human worlds where digitalized technology extends and delocalizes the networks and mediations that circulate reliable witnessing?

Touching Technologies

The reclamation of touch is a wide cultural phenomenon with relevance for ethical speculative considerations. One can just think of how the boom of touch technologies, a market only growing, mobilizes a vast range of more than human reassemblages. How these technologies are made to matter is concomitant with how they transform what matters. Touch technologies emerged in the early 2000s as a promise of what Bill Gates proclaimed to be the "age of digital senses."[6] They "do for the sense of touch what lifelike colour displays and hi-fi sound do for eyes and ears," announced *The Economist* in the early days of haptic hype. The time to lick and sniff keyboards and screens is yet to be trumpeted.[7] For now, technology is "bringing the *neglected* sense of touch into the digital realm."[8] These emerging haptic technologies engaged with a new frontier for the *enhancing* of human experience through computing and digitalized technology. As transhumanist speculations, promises, and expectations about the "innovative" prospects of touch for people in technoscience they constitute a massive matter of investment in a future in which smartphones and other handy devices are only gadget sprouts. Though here I focus on problems posed by the imaginaries of enhancement in everyday experience, the proliferation of applications is vast. Haptic or touch devices are implemented, or fantasized, in relation to many different technologies: for developments of touch sensors in precise industrial robotics[9]; for the creation and manipulation of virtual objects; to allow a feel of materials in video games; to enhance sensorial experience in varied simulators (surgery, sex) and other devices aimed at distant control and operation. They also refer to technologies allowing direct command of laptops and phones through the screen. From the most sophisticated and specialized to the most banal gadgetry, the marketing of these developments uses exciting language that engages play, dexterity of manipulation, augmented or enhanced reality, and experiences of sensorial immersion that mimic *the real thing*, all driven by promises of more immediate connection at the heart of cultural imaginaries of affection. The sense of materiality of contact can take opposed implications; for instance, exposure remains connected to vulnerability so that if it may seem particularly exciting to touch and manipulate "virtual" entities. In other contexts it is reassuring to touch without being touched, to manipulate without physically touching (e.g., in military situations such as the use of drone technology or demining robotics, the viewer remains untouched, touch sensors act as mediators, and distanced bodies and unmanned artifacts receive the immediate physical consequences [Suchman 2016]).

In his essay "Feel the Presence," the haptic geographer Mark Paterson describes these technologies of "touch and distance" and their possibilities of concrete and immediate manipulation of objects, virtual or not. Others and things can be located far away but become "co-present" (Paterson 2006). Paterson explains how adding touch to visual effects produces a sense of "immersion," how these technologies give a feeling of "reality," enhancing the experience of users. However, he shows that the efforts to reproduce and "mimic" tactile sensation are actually productive, performative. An active reconstruction of the sensorial is at stake when developers discuss what will be the *right* feel of a virtual object to implement within the actual design. The transformation of sensorial experience doesn't occur only through *prosthesis* but participates in the "interiorization of technological modes of perceiving" (696; Danius 2002). In other words, touch technologies as more than human assemblages could be remaking what touching means. Inversely, I would add, haptic technology works with the powerful imaginary of touch and its compelling affective power to produce a touching technology, that is, an appealing technology.

Exploring the kinds of more than human worlds that are brought to matter through celebrations of technotouch requires attention to meaning-producing effects emerging in specific configurations. It is not so much a longing for the *real* that is the problem of sociotechnological arrangements that conceal material mediations while pretending quasi-transparent immediacy but rather what will count as real. A politics of care is concerned by which mediations, forms of sustaining life, and problems will be neglected in the count. Which meanings are mobilized—and reinforced—for realizing the promise of touch? By which forms of connection, presence, and relation is technotouch supposed to *enhance* everyday experience? In the technopromises of touch, "more than human" often takes the sense it has for transhumanism, that of a desire to transcend human limitations. A trend that, far from decentering human agency via a more than human reassemblage, reinforces it even if disembodied, aiming at making humans more powerful through technoscientific progress. As the protagonist of David Brin's SF novel puts it, as he collects trash from space with an extended body that connects his isolated, encapsulated, imperfect body to a distant outer space, a "more real" world is the dream:

> The illusion felt perfect, at last. . . . Thirty kilometers of slender, conducting filament.
>
> . . . At both ends of the pivoting tether were compact clusters of sensors (my eyes), cathode emitters (my muscles), and grabbers (my clutching hands), that felt more part of him, right now, than anything made of flesh. More real than the meaty parts he had been born with, now drifting in a cocoon far below, near the bulky, pitted space station. That distant human body seemed almost imaginary.

Dreams of technological extension beg a more specific question: Which qualities are selected for human improvement? The question of enhancement does not need us to examine any particularly extravagant science-fiction scenarios; it is visible in the most ordinary settings. In the early days of excited hype about haptic technology, *tactile technologies*, a company dedicated to the development and expansion of touch screens, advertised the benefits in its promotional website.[10] The first claimed advantage was speed: "Fast, faster, fastest." Touch screens cut time waste through direct touch in a world where "being one second faster could make all the difference." This directness is enhanced and integrated for "everybody," as a second advantage is promoted: "touch makes everybody an expert" by "intuitive" reaching out; "you just point at what you want." To touch is to get. Expertise would ameliorate as "touchscreen-based systems virtually eliminate errors as users select from clearly defined menus." The goal is intuitive immediacy, reduction of training to *direct* expertise, elimination of mistakes based on preordered selection. In conclusion, they offer a "naturally easy interface to use" for what the job requires: efficacy and speediness, reduction of training time, and keeping costs down. On top of these advantages—hands being guilty vehicles of everyday contagions—touch screens are purportedly "cleaner." This company therefore offered systems that are "not affected by dirt, dust grease or liquids." Here the driving dream is not so much of enhanced reality but enhanced effectiveness and speed. Touch stands for unmediated directness of manipulation, while hygiene worries respond to remnants of involved flesh. This is a particular vision of the more than human reassemblage offered by touch technologies, one that rather than innovating relation reinforces prevalent conceptions of efficiency—identified to accelerated productiveness. In the last chapter of the book, I will engage with how the paradigm of productivity, accelerated speed, and focus on output affects the temporality of care. What the ambivalent value of touch exposes here is that enhancing material connection does not necessarily mean awareness of embodied effects.

Computers are touching technologies in a very special way via keyboards, screens, and mouses. As somebody who spends a great amount of time behind a computer, I am not immune to the seductive hype of smooth touch screens. But as an intermittent member of the community affected

by Repetitive Stress Syndrome and other health hazards of the computer-
ized workplace, I also wonder why possible innovations offered by these
technologies for at least not worsening this epidemic are not being promoted. Many users' computing experience includes diverse ergonomic
devices that make repetitive touch labor easier and dress up the cyborg
imaginary of flesh wired to a keyboard (adapted mouse and keyboard,
wrist and back elastic bands, microphones and voice recognition software,
etc.). In order to situate keyboard-related illness as a historically collective
phenomenon, it is insightful to read Sarah Lochlann Jain's account of the
injury production concomitant to this device's history. Making touching
technologies a matter of care requires that we learn about the possibilities
overlooked by an industry in hasty development: missed opportunities to
be in touch with the consequences that constant keyboard touch feedback
doubled with pressures of efficiency has had on user's everyday lives (Jain
2006). Touch and proximity belong to the conceptual nebula of care, but
they are not caring per se.

   And yet yearnings of proximity in caring involvements mark the everydayness of computing technology. These are finely expressed in a poem by
Susan Leigh Star, who also raises ambivalent feelings about promises of
enhancement via technical extension:

ii
my best friend lives two thousand miles away
and every day
my fingertips bleed distilled intimacy
trapped Pavlovas
dance, I curse, dance
bring her to me
the bandwith of her smell

ii
years ago I lay twisted
below the terminal
the keyboard my only hope
for work

for continuity
my stubborn shoulders
my ruined spine
my aching arms
suspended above my head
soft green letters
reflect back
Chapter One:
no one can see you
Chapter Two:
your body is filtered here
Chapter Three: you are not alone (Star 1995, 30–31)

Computers are more than working prostheses; they are existential companions for people trying to keep in touch with dislocated networks of loved ones. *My sister lives ten thousand miles away*—my parents, siblings, and friends are spread throughout the World Wide Web. A scattered heart, bleeding fingertips, and a ruined back, frustrations of "distilled intimacy," are not enough to stop efforts to remain in touch through screens. E-political communities in a globalized world also depend on virtual touching and social media props. Haptic technologies feel particularly appealing for those for whom mobility has transformed community and who have to "survive in the diaspora" (Haraway 1991a, 171). Touch technologies and longings of being in touch match well. The remaking of sensorial experience through the intensification of digital touch feeds on the marketing of proximities in the distance and our investment in longing.

Yearnings for touch, for being in touch, are also at the heart of caring involvement. But there is no point in idealizing the possibilities. If touch extends, it is also because it is a reminder of finitude (why would infinite beings yearn for extension?). And if touch deprivation is a serious issue, *overwhelming* is the word that comes to my mind when enhancement of experience is put at the forefront. Permanent *intouchness*? With what? Like care, touch is not a harmless affection. Touch receptors, located all over our bodies, are also pain receptors; they register what happens through our surface and send signals of pain and pleasure. When absorbed by work and e-relations, these sensations take time to be perceived. We can get relatively out of touch with what bodies endure and forget the care and labor that is needed to get them through the day. There is no production of virtual relationality, whether commodified by capitalist investment or consumer society, that will not draw upon the life of some-body somewhere. Kalindi Vora shows, for instance, how the "vital energy" of call-center workers in India is drained by the overnight labor required for keeping in touch with the needs of clients in North America to which their bodies are invisible in turn (Vora 2009b). Insisting on the many ways in which digitalized technologies engage material touching of finite flesh renders insufficient the qualification of knowledge economies and affective labors as "immaterial" (Hardt and Negri 2000, 290). More alertness to chains of touch in digital culture could also expand awareness of the layers of material mediations that allow technological connection. Besides human labor, virtual technocultures always touch some*thing* somewhere—through demands for electric power generation and the proliferation of high-tech trash (Stephenson 1996; Basel Action Network 2002; Strand 2008).[11]

As I have argued above, transforming purported facts and objects into matters of care by thinking with and for neglected labors and marginalized experiences is a way to remain in touch with problems erased or silenced by thriving technoscientific mobilizations. This means addressing innovatory technologies that are supposed to enhance living conditions with questions about the social relations, labors, and desires that may become obliterated through their development, use, and implementation. Such issues appear particularly relevant in another field of haptic research investment and expectations to enhance ordinary experiences. I am thinking of distant surgery where touch sensors seek dexterity in distant manipulation (Satava 2004). The rationale here is not *more* touching but improving the chain of technological mediations in order to give a sense of directness and precision of touch while accessing distant flesh and bodies. The surgeon could become physically absent, a "telepresence" that, however, can work simultaneously on multiple patients. A possible reduction in number of nurses that will do the work *on site* is also invoked. Again we encounter "the epitome of efficiency," understood purely in quantitative terms: reduction of costs and human resources. If complex chirurgical intervention is not yet realizable this way, healing through telecare is not a fantasy. Sometimes it aims to enhance access to health care in deprived locations where developing haptic technologies for co-presence makes sense. However, we need also to ask what types of experiences of caring will be produced through these innovations? Which new managed "conducts" will pass as care? (Latimer 2000). Thinking from labors that become less visible and from the perspective of patients/users and, importantly, also that of "non users," Nelly Oudshoorn shows how care at a distance challenges existing modes of interaction and *transforms* rather than reduces burdens of labor. Also, the replacing of face-to-face interaction places sections of the networks of health care out of touch for patients (Oudshoorn 2008a; 2008b). The materiality and directness of touch acquires added tones as other mediations are rendered irrelevant: What are more efficient doctors going to be in touch with? What kind of healing-touch is this? Is the reversibility of touch, its potential of consequential corelationality, of shared vulnerability, invalidated when patients cannot reach who is touching them?[12] One thing seems sure in a finite world, that these new forms of connection produce as much copresence as they increase absence. They do not really *reduce* distance; they redistribute it.

Pausing: Dilemmas of Speculative Thinking

Questions and skepticism about expanded possibility in promises of touch accumulate. Yet my aim is not to distance myself from these yearnings, neither to purify an "other" vision of touch—the "really" caring one. I am not interested in the elucidation of underlying social, political, and cultural reasons and causes for the lure of touch and the attractiveness of promises of technotouching. I could be discussing how this "turn" to touch may correspond to other declared theoretical turns: turns to materiality, to practices, to ontology, to radical empiricism. But while I am hesitating here about the promises of touch, I remain concerned about the pitfalls of theoretical critique discussed in the previous chapters. Blanketing the specificities of situations and cases under a general rationale that critiques the haptic promise, placing myself as observer at a distance from where I could understand what is at stake, would be falling into one of those pitfalls. Zooming out at theoretical speed, blending categories that mirror each other into a feel of sameness to support the argument that *something is happening* in the turn to touch might be precisely what thinking with touch, thinking haptically, is not about: the specificity of textures disappears and "a" problem surreptitiously becomes everybody's problem.

My engagement with touch remains situated within an exploration of what caring signifies for thinking and knowing in more than human worlds. Here, a caring politics of speculative thinking could reclaim *hapticity* as a way to keep close to an engagement to respond to what a problem "requires." And of course, what we come to consider problematic is grounded in the collective commitments that shape our thinking and what we care for. And yet a speculative commitment grounded in the problems that we have set out to respond to seeks not to "simply reflect that which, a priori, we define as plausible" (Stengers 2004), or that which confirms a theory. In other words, engaged speculative responses are situated by what appears as a problem within specific commitments and inheritances, within contingencies and experiences in situation. If to care is to become susceptible of being affected by some matters rather than others, then situated responses are engaged in interdependent more-than-one modes of subjectivity and political consciousness. Therefore, in revaluations of touch, in reclamations of touch, not only do I read the kind of world-making that is being speculated upon through the partialities of my cares but I also think with other speculative possibilities.

That things could be different is the impulse of speculative thinking. In this book the speculative refers to a mode of thought committed to foster visions of other worlds possible, to paraphrase the motto of the alter-globalization movement, "another world is possible."[13] Related to the sense of sight, the way of the speculative is traditionally associated with vision, observation. In feminist approaches, as I mentioned in the introduction, speculative thinking fuels hope and the desire for transformative action. It belongs to feminism visions' affective power to touch, to nurture hope about what the world could be, and to engage with its promises and threats (Haran 2001). This involves political imagination of the possible, purposes of making a difference with awareness and responsibility for consequences: speculative thinking as involved intervention—as speculative commitment.

But the notion of speculative vision also seems to suggest—as in the phrasing "pure speculation"—a flight transcending the material conditions that ground transformation in the present, from the plainness and mundaneness of the everyday that visionaries are habitually suspected of neglecting. The predicament of speculative thought somehow reenacts a worn-out fraught question for critical thought: How can thinking lead to material change? And paradoxically, it doesn't help that vision, as a metaphor for knowing, has traditionally conveyed the notion that true thought and knowledge is based on clear and unpolluted observation and reason, on a disembodied relation to a distinct world, the pride of modern science according to rationalist humanist philosophies. If the speculative is suspected of improbability, thought and action led by metaphors of clear vision have been criticized for a reductionist, bifurcated, form of relating, abstracted from the bodily engagement that makes knowing subjects relevant in interdependent worlds. What's more, opting for the speculative as the making of a difference, for diffraction rather than reflection of the same, for alternative investments in thinking the possible or the virtual, I also have to consider my belonging to a time and culture radically turned into investment into a future (of outputs and returns of investment) in ways that tend to drain present everyday conditions (an issue that I address in the last chapter of the book). In my world, the speculative is also the name of fairly intoxicating financialized bubbles out of touch with finite pasts, presents, and futures. These unsolved tensions are embedded in an attempt of thinking with care invested in speculative thinking of what could be but grounded in the mundane possible, in a hands-on doing connected with neglected everydayness.

Devising relevant and grounded interventions calls for speculative thinking that goes beyond descriptions and explanations of what is and of how things came to be. The worlds into which touch will attract us are not written in its technologies or in the purported nature of touch's singular phenomenology. The concrete differences made when reclaiming touch and reinventing touching technologies for everyday life are all but neutral; they will be marked by visions that touch us, and that we want others to be touched by, speculative visions of touch—touching visions. Where this consideration of the ambivalent promise of touch for thinking speculatively with care has brought me is to questions such as: How can visionary diffractive efforts resist inflated virtual (future) possibility detached from (present) material finitudes? And can we resist the promises of immanent touch to transcend fraught mediations?

Touching Visions

My initial leaning for touch as a sensorial universe that expresses the ambivalences of caring emerged from its potential for responding to the abstract and disengaged distances more easily associated with knowledge-as-vision. But because touch short-circuits distance, it is also susceptible to convey other powerful expectations: immediacy as authentic connection to the *real*, including otherworldly realities for spiritual or mystic traditions, as well as claims not so much of transparent and unpolluted observation but of *direct* and extended accelerated *efficient* intervention. If touch could offer a sensorial, embodied grounding for the proximities of caring knowing, we also need touching visions more susceptible to foster accountability for the mediations, ambivalences, and eventual pitfalls of touch and its technologies. Connected bodily experience is not per se oriented to improve caring, nor does reducing distance necessarily trouble in predominant oppressive configurations. It is in this spirit that I return now to interventions that engage with touch to reclaim vision, by manifesting deep attention to materiality and embodiment in ways that rethink relationality, in ways that suggest a desire for tangible engagements with mundane transformation.

A grounded vision of transformation, rather than "enhancement," of experience through touch can be read in how Claudia Castañeda engages speculatively with the "future of touch," exploring specific touch-abilities in developments of "robotic skin" (Castañeda 2001). One of the stories she critically engages with is that of a "bush" robot constructed with a trillion tiny "leaves," each equipped with tactile sensors. This touchy leafy skin would, according to its conceiver's ambitious vision, see *better* than the human eye, for instance, by feeling a photograph or a movie through directly touching its material (227). Castañeda is interested in the "suggestiveness" of such a robotic formation for feminist theories of embodiment and relationality: "What would it be like to touch the visual in the way this [robot] can?" Castañeda argues that when vision is "rematerialized" through direct contact, refusing the distinction between vision and touch troubles the ground of objectivity: "the distinction between distanced (objective) vision and the subjective, embodied contact" (229). Yet her vision of touching futures doesn't translate in a promise of overcoming (human) limitations. On the contrary, Castañeda reminds us that robotic touch is not limitless; it responds to the technological reproduction of specific understandings of how touch works.

In other projects Castañeda looks for alternatives, where robotic skin is rather conceived as a site of learning in interaction with the environment. One characteristic of these learning robots' interactive skin is that it first acts as protection: an alarm system that assists in learning to distinguish what is harmful and can destroy it (Castañeda 2001, 231). The requirement and outcome of ongoing technohaptic learning is not here mastery of dexterous manipulation but a skillful recognition of vulnerability. This suggests that, in contrast with dreams of directness, implementing touching technologies could foster awareness that learning (to) touch is a process. Developing skills is required for precise and careful touching, for learning *how to touch*, specifically. The experience of touch can then serve to insist on the specificity of contact. Castañeda draws from Merleau-Ponty to argue that the experience of touch "cannot be detached from its embodiment," but neither is it "reducible to the body itself." The skin, as an active living surface, "becomes a site of possibility" (232–34). In this vision, the generative character of touch is not given; it emerges from contact with *a* world, a process through which a body learns, evolves, and becomes. All but a dream of immediacy. The affirmation of specificity of contact and encounters is also not a limitation imposed on possibility. Specificity *is* what produces diversity: this is precisely how touch can have multiplying effects, extending the range of experiences rather than extending one mode of experience.

We can go further to affirm that touch is world-making, a thought that resonates with the relational ontology for which being *is* relating approached in the previous chapter. We can read Karen Barad's (2007) account of the seeing-touching made possible by "scanning tunnelling microscopes" in this direction. These devices are used to "observe" surfaces at atomic level, a procedure that operates "on very different physical principles than visual sight" (53). This account calls upon the "physicality of touch." A sense of the object passes through a "microscope tip" and the "feel" of the surface passes through an electron current tunneled through the microscope. The data produced (including the resulting image of the surface) corresponds to "specific arrangements of atoms." In this encounter, where the physical universe is as much an agent in the *meeting* with a knower, there is no separateness between observing and touching, figuring well a vision that does not separate knowing from being-relating. Barad's account of the closeness of touch stands for a conception where "knowing does not come from standing at a distance and representing the world but rather from *a direct material engagement with the world*" (49, emphasis added).

This vision challenges the framing of knowing within epistemologies of representation and "optics of mediation" (Barad 2007, 374–77)—in social constructivism, for instance, "nature" never comes to "us" but is mediated by the knowledge social beings have of it. A critique of this bifurcated optic order requires a more subtle thinking of the "agency" involved in knowing yet without necessarily speaking for immediacy, for directness in touching the real, or nature. On the contrary, vision-as-touch works rather to increase a sense of the entanglement of multiple materialities, as in Barad's theory of the "intra-activity" of human and nonhuman matters in the scientific constitution of phenomena. Going further than inter-action, Barad's intra-action problematizes not only subjectivity but also the attribution of agency merely to human subjects (of science)—as the ones having power to intervene and transform (construct) reality. The reversibility of touch (to touch is to be touched) also inspires the troubling of such assumptions: Who/what is *object*? Who/what is *subject*? It is not only the experimenter/observer/human agent who sees, touches, knows, intervenes, and manipulates the universe: there is *intra-touching*. In the example above, it is not only the microscope that touches a surface; this surface *does* something to the artifact of touching-vision. In other words, touching technologies are material and meaning producing embodied practices entangled with the very matter of relating-being. As such, they cannot be about touch and get, or about immediate access to more reality. Reality *is* a process of intra-active touch. Interdependency is intrarelational.

As it undermines the grounds of the invulnerable, untouched position of the master subject-agent that appropriates inanimate worlds, this ontology carries ethical resonance. What we do in, to, a world can come back, re-affect someone somehow. This is thinking touch as world-making. How we know in the world populates it with specific connections. People and things "are in mutually constituting active touch" that "rich naturecultural contact zones multiply with each tactile look" (Haraway 2007b, 6–7). Thought as a material embodied relation that holds worlds together, touch intensifies awareness about the transformative character of contact, including visual contact— tactile looks. Here the sense of intensified curiosity is figured by a particular way of seeing-touching, a haptic-optic figured by Eva Hayward's "fingeryeyes." Coined in speculative thinking with the sensorial impressions of encountering cup corals, this figuration speaks of a visual-haptic-sensorial apparatus of "tentacular visuality" as well as the "synaesthetic quality of materialized sensation" (Hayward 2010, 580). Hayward's sensuous writing compels us into the queerness of caressing encounters with cup corals but retains awareness of the predicaments of closeness to fragile nonhuman others:

> The coralogical impressions of fingeryeyes that I have described cannot be agnostic about animal well-being because ontology is what is at stake. Cross-species sensations are always mediated by power that leaves impressions, which leaves bodies imprinted and furrowed with consequences. Animal bodies—the coral's and mine—carry forms of domination, communion, and activation into the folds of being. As we look for multispecies manifestations we must not ignore the repercussions that these unions have for all actors. In the effort to touch corals, to make sense of their biomechanics, I have also aided in the death of the corals I describe here; this species-sensing is not easily refused by the animals. (592)

What these visions that play with vision as touch and touch as vision invite to think is a world constantly done and undone through encounters that accentuate both the attraction of closeness as well as awareness of alterity. And so, marked by unexpectedness, they require a situated ethicality.

There is a particular form of multifaceted collective reciprocity at stake in the ability and responsibility to respond to being touched: a "response-ability," in Haraway's terms. This requires curiosity about what happens in contact zones, asking question such as: "whom and what do I touch when I touch my dog?" with which Haraway opens her adventurous exploration of the layers of naturecultural relations that make interspecies touchings possible—including sophisticated and mundane technologies—while actively speculating on what could be possible through taking seriously these chains of touch. These are worlds of collective feeling, relational processes that are far from being always pleasant or livable but have something specific and situated to teach us. The question of how we learn to live with others, being in the world—to be touched as much as to actively touch, is an opening to "becoming with." Touch "ramifies and shapes accountability" (Haraway 2007b, 36), furthers a sense of inheriting "in the flesh," and invites us to be more aware about how living-as-relating engages both "pleasure and obligation" (7). In contrast with promises of touching technologies for network extension and human enhancement thinking about caring proximities, these situated touching visions can increase ethical awareness about material consequences. Here, knowing practices engage in adding relation to a world by involvement in touching and being touched by what we "observe." Thinking with these visions, I seek a sense of touch that doesn't evoke a hold on reality with improved grasp that intensifies proximity with gradualness and care, attention to detail in encounters, reciprocal exposure, and vulnerability, rather than speeded efficacy of appropriation.[14]

A beautiful example of a nuanced reclamation of touch, paradoxically within a reaffirmation of vision, is how, in her analysis of close-up images, taken at an almost touching closeness, media theorist Laura U. Marks describes the blurred figures produced by intimate detailed images of tiny things, inviting the viewer into "a small caressing gaze" on pores and textures at the surface (Marks 2002, xi). She argues that the power of a *haptic* image is not the identification of/with a distinct "figure" but to engage viewer and image in an immersed "bodily relationship." Yet wanting to "warm up" rather than negate optic culture, Marks doesn't aim to abolish distance but rather to keep an "erotic oscillation" in which the desire of banishing distances is in tension with the letting go of the other, not driven by possessiveness (13–15). Significantly, she says that the closeness of haptic visuality induces us to acknowledge the *"unknowability* of the other." When vision is blurred in close imagery, objects become "too close to be seen properly," "optical resources fail to see," and optic knowing is "frustrated." It is then that the impulse of haptic visuality is stirred up, inviting us to "haptic speculation" (16). We learn that to speculate is also to admit that we do not *really* know wholly. Though there are indeed many things that knowledge-as-distant vision fails to feel, if touch augments proximity, it also can disrupt and challenge the idealization of longings for closeness and, more specifically, of superior knowledge in proximity.

Haptic speculation doesn't guarantee material certainty; touching is not a promise of enhanced contact with "reality" but rather an invitation to participate in its ongoing redoing and to be redone in the process. Dimitris Papadopoulos, Niamh Stephenson, and Vassilis Tsianos (2008b, 143) conceive a haptic approach to engage with transformative possibilities in everyday forms of sociability that are neglected by optic representation. They encourage haptic experiencing as an attempt to change our perception, to "hone" it to perceive the "imperceptible politics" in everyday practices in which another world is *here*, in the making, before "events" become visible to representation. In these they see a chance, not only for subversion but for creating alternative knowledges. Haptic (political) experience is for them a craft of *carving* possibility in the midst of potential incommensurability. Unknowability takes here yet another meaning.[15] Haptic speculation is not about imaginative expectation of events to come; it is the everyday (survival) strategy rooted in the present of "life below the radars" of optic orders that do not welcome, know, or not even *perceive* the practices that exceed preexistent representations and meanings. It is not difficult to see why this way of being-knowing with a world can be attuned to the sensibilities of thinking with care, to honing perception to matters of care. Focusing on everydayness, on the uneventful, is a way of noticing care's ordinary doings, the domestic unimpressive ways in which we get through the day, without which no event would be possible. While events are those breaks that make a difference, marking a before and after that gets recorded in history, care, in spite of all the work of political reclaiming, in spite of its hegemonic marketization, remains associated with the unexciting, blended with the dullness of the everyday, with an uneventful temporality. Haptic engagement is akin to thinking with care as a (knowledge) politics of inhabiting the potentials of neglected perception, of speculative commitments that are about relating with, and partaking in, worlds struggling to make their other visions not so much *visible* but possible. These engagements do not so much entail that knowing will be enhanced, more given, or immediate through touch than through seeing; rather, they call attention to the dimension of knowing, which is not about elucidating, but about affecting, touching and being touched, for better or for worse. About involved knowing, knowledge that cares.

Coda: Sensory Values

> Kira laid a slim hand on the bulkhead, on the square plate that was the only access to Helva's titanium shell within the column. It was a gesture of apology and entreaty, simple and swift. Had Helva been aware of *sensory values* it would have been the lightest of pressures. (McCaffrey 1991, 35, emphasis added)

Kira is a human traveling through space *in* Helva, a female-gendered spaceship with a human brain, the central character of Anne McCaffrey's science-fiction classic, *The Ship Who Sang*. These two beings are starting their first conjoint mission and learning to know each other. Both are touchy, in intense pain due to the loss of loved ones (a husband in Kira's case, the previous human ship skipper in Helva's). The excerpt above comes from a scene where Helva, the ship, is physically touched by Kira after a moment of tense argument between them. Helva has no skin sensitive to "sensory values"; however, she indeed *feels* something, beyond her titanium shell body, just by seeing Kira's touching gesture. Helva cannot touch Kira back; her power to act through physical touch is limited. She touches Kira through careful word communication, and by readjusting functions in order to create a caring environment for her in her body-spaceship. Kira knows that Helva's titanium shell cannot "feel" her touch and still her gesture of apology expresses the "lightest of pressures," which Anne McCaffrey qualifies as a "sensory value."

Throughout this chapter I have used "vision," instead of sight, to refer to visual sensorial universes and to speculative ethico-political imagination. Lacking a word that makes of touch what vision makes of sight, I have used touching visions as a surrogate. The promise of touching visions is not just given by the haptic's particular phenomenology. Following the lure of the haptic, I ended up looking for visions that could engage touch with care, that is, that do not idealize it. Without proposing these to become normative orientations, I wonder what it could mean to foster something like "sensory values" for the power of touch, for our touching technologies? I'm thinking of values as collective ventures embodied and embedded in prosaic material everyday agencies, contingently becoming vital to situated relationalities that ground them in a living web of care; of values not necessarily as that which should define the good but as interrogative demands emerging from relations. Sensory values are not qualities reserved to touch, but thinking with touch emphasizes them well because of the intensification of closeness that the haptic signifies and enacts. Touching technologies do not need to celebrate the inherent significance of touch but rather touching visions that also account for haptic asperities. Values for touching visions call for an ethical engagement with the possibility of care as a relation that short-circuits (critical) distance and that is about immersed, impure, ethical involvement, but remain in tension with both moral orderings—such as managerial orientations toward efficiency and speed—and idealized longings for immanent relations.

A sensory value in Kira and Helva's interaction inspired by the trope of touch could be named "tactfulness," the same word for the sense of touch in some languages—for example, in Spanish, *tacto*. A form of sensorial politeness, understood as a political art of gauging distance and proximity.[16] An ethical and political learning that might well be vital in caring for worlds in the making through intensified, constant touch between entities human and more than human—a daily practice of "articulating bodies to other bodies with care so that significant others can flourish" (Haraway 2007b, 92). Thinking touch with care beautifully emphasizes intra-active reversibility, and therefore vulnerability in relational ontologies. If touch is an experience where boundaries of self and other tend to blur, it also speaks of intrusiveness and appropriation: it *is* possible to touch without being touched. Appropriation abolishes significance. Thought through a politics of care, "intra-active" touch demands attentiveness to the response, or reaction, of the touched. It demands to question when and how we shall avoid touch, to remain open for our haptic speculations to be cut short by the resistance of an "other," to be frustrated by the encounter of another way of touching/knowing. A sense of careful "reciprocity" could therefore be another value for thinking with touch's remarkable quality of reversibility.

Thinking sensory values of care with the universe of touch is a speculative displacement of ethical questioning. Reciprocity is an interesting notion to expose this. Thinking the webs of care through sensorial materiality, as chains of touch that link and remake worlds, troubles not only longings for closeness but also the reduction of relations of reciprocity to logics of exchange between individuals. Sensory values such as intra-touching politeness and haptic reciprocity refer to an obligation to reciprocate attentiveness to others, but one that is quite different from that of a moral contract or the enactment of norms—a quality of caring obligations that I discuss in the next chapter. Thinking care through the haptic and the haptic through care brings up one of the most appealing aspects of care for a speculative ethics in more than human worlds: that its "value" is inseparable from the implication of the carer in a doing that affects her. Care *obliges* in ways embedded in everyday doings and agencies; it obliges because it is inherent to relations of interdependency.

Affirming care as an inherently material obligation is a fraught terrain, given what this means for caregivers, that caring is often a trap, a reason why, as Carol Gould has argued, reducing political obligation to consent or choice is an extremely gendered ideal that excludes a whole set of relations from the political sphere where choice and consent between autonomous individuals has little meaning (Gould 1988). Here I am obviously arguing for a distributed notion of the material obligation of care—not as something that only some should be forced to fulfill.[17] Thinking reciprocity through a collective web of obligations, rather than individual commitments, exposes the multilateral circulation of agencies of care.[18] As David Schmidtz argues, the common idea of "symmetrical" reciprocity doesn't exhaust the ways people try to "pass on" a good received (Schmidtz 2006, 82–83). Care troubles reciprocity in this way because the living web of care is not one where every giving involves taking, nor every taking will involve giving. The care that touches me today and sustains me might never be given back (by me or others) to those who generated it, who might not even need or want my care. In turn, the care I will give will touch beings who never will give me (back) this care. Reasons to support this vision are advanced by work that sees the ethical implications of care challenging an ethics based on "justice" (Gilligan 1982). And why others ask for the reciprocity of care to be collectively distributed (Kittay 1999), contest the reciprocity model of economic exchange, support "unconditional welfare" (Segall 2005) for example, the State would provide means for care (through unconditional basic income) that could ensure that those with care responsibilities, but who might not have somebody caring for them, are not depleted or neglected. And so by being cared for, they also continue to be able to care for others. Whether we agree or not that the state, given its major role in the structural reproduction of inequalities, is the appropriate collective to foster an ethics inherent to communally reciprocal relations, the essential notion here is that reciprocity in *as well as possible* care circulates multilaterally, collectively: it is shared. Iris Marion Young adds another problematic dimension to these relations when she argues that reciprocity cannot be thought as symmetrical because this masks the asymmetrical positions in which people are situated and the possibility of a different ethics: "opening up to the other person is always a *gift*; the trust to communicate cannot await the other person's promise to reciprocate" (Young 1997, 352). I propose to think of relations of care giving and receiving in a similar way not so much because care is a gift but because there is no guarantee that care will be reciprocated; it happens asymmetrically both in terms of power and because people who care, caregivers, cannot give with the expectation for it to be symmetrically reciprocated. The care that has been "passed on"—as is neglect—continues to circulate, not necessarily morally or intentionally, in an embodied way, or simply embedded in the world, environments, infrastructures that have been marked by that care. The passing on of "care" does not need to be determined by the care we have received to be tangible. What these multilateral reciprocities of care disrupt are conceptions of the ethical as a moral compound of obligations and responsibilities presiding over the agency of intentional (human) moral subjects.

In the following chapters, we will see how these questions have brought this journey closer to attempts to think differently about the circulation of ethicality in more than human worlds—close to those who contest the reduction of ethicality to human intentionality (Barad 2007) and to those who engage with the intentionality of the other than human, seeking to think of "nature in the active voice" (Plumwood 2001). These are paths for questioning human-centered notions of agency that do not necessarily converge, but they are both compelling and challenging to thinking with care in more than human worlds. Interrogating the intra-active but non-bilateral reciprocity of touching with care for the touched, thinking touch through care and as sensory values, invites us to distribute and transfer ethicality through multilateral asymmetrical agencies that don't follow uni-directional patterns of individual intentionality. Caring, or not caring, however, are ethico-political problems and agencies that we mostly think as they pass *from* humans toward others. But thinking care with things and objects exposes that the thick relational complexity of the intratouching circulation of care might be even more intense when we take into account that our worlds are more than human: the agencies at stake multiply. How to care becomes a particularly poignant question in times when other than humans seem to be utterly appropriated in the networks of (some) Anthropos. What does it mean to think how, in the web of care, other than humans constantly "reciprocate"? Can we, at least speculatively, include such thoughts in an ethical inquiry modestly reaching out with care from the uneasy inheritances of human antiecological situatedness? Following such intimations, Part II of this book attempts to think care as a generalized condition that circulates through the stuff and substance of the world, as agencies without which nothing that has any relation to humans would live well, whether all that is alive is engaged in giving or care, whether care is intentionally ethical.

PART II

Speculative Ethics in Antiecological Times

While we may all ultimately be connected to one another, the specificity and proximity of connections matters—who we are bound up with and in what ways. Life and death happen inside these relationships.

—THOM VAN DOOREN, *Flight Ways*

Crucially, there is no getting away from ethics on this account of mattering. Ethics is an integral part of the diffraction (ongoing differentiating) patterns of worlding, not a superimposing of human values onto the ontology of the world (as if "fact" and "value" were radically other). The very nature of matter entails an exposure to the Other. Responsibility is not an obligation that the subject chooses but rather an incarnate relation that precedes the intentionality of consciousness. Responsibility is not a calculation to be performed. It is a relation always already integral to the world's ongoing intraactive becoming and not-becoming.

—KAREN BARAD, *Quantum Entanglements and Hauntological Relations of Inheritance*

four

≈≈

Alterbiopolitics

Care of earth, care of people, return of the surplus.

—Earth Activist Training, *Principles of Permaculture Ethics*

In May 2006, I traveled to the hills that overlook Bodega Bay, an hour and a half north of San Francisco, to participate in two weeks of intensive training in permaculture technologies. The course was organized by the U.S.-based Earth Activist Training (EAT) collective, and the lead teachers were Eric Ohlsen, a permaculture expert practitioner, ecological landscape designer, and activist, and Starhawk, a renowned pagan spiritual figure, writer, and activist.[1] With around thirty other participants from different backgrounds of practice—organic farmers, city growers, ecological activists, community organizers, engineers, forest managers—I was introduced to permaculture technologies for ecological practice as a form of tangible activism based on a commitment to care for the earth. The Earth Activist Training collective linked its teachings to this particular version of a triple motto—"care of earth," "care of people," "return of the surplus"—that circulates in permaculture networks as the principles of permaculture ethics (another version being "Earthcare, Peoplecare, Fairshares" [Burnett 2008]).

Until now, this book has engaged with care's potential to open ethico-political questions in research and thought concerned by the consequences of scientific and technological reconfigurations of more than human worlds. I've been looking out for ways of thinking about and with care that do not correspond with normative moral and epistemological orders but that could displace them. The chapters in Part II continue this journey through terrains opened by my experiential immersion into ethico-political reconfigurations of ecological relations. This journey started in an encounter with the practices and movements of permaculture and led me to research human–soil relations, which I discuss in the last chapter. The work that follows is therefore grounded in terrains, involved with those worried about antiecological times, processes, and agencies. But though the thinking here connects with more substantial accounts of specific landscapes of care, it remains driven by a speculative search for critical stories that feed a sense of possibility. It is in this spirit that I engage first with fragments of my experience of encountering permaculture, especially foregrounding the orientations it gives to think the "ethicality" in care.

Permaculture is a global movement with many local actualizations of which my knowledge is partial, starting from a particular collective.[2] I have to say that I didn't initially feel compelled to write about my experience and relation with this movement. Quite the contrary. I didn't see this experience in any way connected to my academic work. It was only years after this initial training that I started realizing that it had already permeated my thinking. I was intrigued by how care figured in permaculture statements as a principled ethics of care.[3] I felt compelled to ask: To which kind of ethics do these principles belong? Why had I felt them as transformative and hopeful rather than constraining? In EAT trainings, the teachings were not about morality; nor did we spend much time discussing ethical implications. The focus was on learning how to make and live with everyday systems and techniques that embody and embed care for the earth. Attempting to think these ethics brought me a deeper understanding of care as a politics and an ethics concomitant to the everyday materialities of life. It also required closer thought on the displacements of care in an ethics concerned with redoing relations in more than human living webs.

While permaculture has gained organizational identity, many of the techniques and practices it promotes are not exclusive to this label—they are shared with and/or borrowed from agroecology, biodynamical agriculture, indigenous modes of land care, and more. Permaculture ethics therefore offers a window into related spheres of alternative ecological doings. Generally speaking, the permaculture movement is known to promote technologies that foster ecological living (urban and rural) through the design of alternative systems for local food production, waste management, re-source renewal, alternative energies, and radical democratic forms of orga-nization. Permaculture practices have extended through practice-sharing, teaching, community building, and ecosocial activism, as well as through the published work, theoretical and practical, of important figures such as David Holmgren and Bill Mollison and of a large community of research-ers and practitioners. Across writings, practices, and interventions, its pro-ponents envision the ethical and political effectiveness of permaculture in the possibility of transforming people's ways of going about our *everyday* relations to the earth, its inhabitants and "resources." This vision was at the heart of the training I followed. In word and practice this is an ethics embedded in concrete mundane relationalities. Principles are inseparable from practices at the level of ordinary life. This means, as I explore in this chapter, that *personal* practice and "private" forms of living are connected to a collective in an intrinsic way and that it is *ethos* that grounds ethical principles rather than follows them. As ethical obligations and commit-ments that do not start from a normative morality or from an individual-ized subject, these notions are inspiring for a speculative exploration of ethical involvement.

Another reason that connects this way of conceiving everyday ecologi-cal practices as ethics to a speculative way of conceiving care is the engage-ment with the tasks and consequences of living in naturecultures. This approach to human relations to the nonhuman world and forces cannot be easily described along the lines of a humanist/posthumanist binary. Rather than taxonomizing this movement along traditional bifurcations, I read permaculture as a timely intervention at the heart of the contemporary awareness that we live in a *naturecultural* world. Naturecultures, as used by Haraway (1997b), signify the inseparability of the natural and the cultural in technoscience and a rejection of humanist ontological splits in modern traditions (Haraway 1991a; see also Latour 1993). This is not an evident reading, I admit. One of the many manuals of permaculture defines it like this: "Permaculture is about creating sustainable human habitats by *follow-ing nature's patterns*" (Burnett 2008, 8, emphasis added). The orientation to follow nature's patterns is widespread, but it would be a mistake to reduce it to an antitechnological or a back-to-untouched-natural state. Though this tendency also insists in this large and multifarious movement, most practitioners embrace permaculture as the search for alternative technologies that work with natural mechanisms rather than against them (in the next chapter I read this stance as a refutation of reductive notions of innovation and the charge of a regressive temporality). This is not just biomimicry proper. As the renowned U.S. permaculturist Penny Livingston puts it, the issue is not so much for humans to act upon the "environment" but to consider that "we are nature working" (cited in Starhawk 2004, 9). And yet the term "permaculture" itself—usually attributed to David Holmgren and Bill Mollison (see Holmgren 2002)—puts "culture" at the forefront, indicating also the purpose of *cultivating* ongoing communal practices over time (acting within a community of human and nonhuman beings) that foster a certain durability of (permanent) renewal and fruitfulness versus the antiecological depletion of resources. The natural and the cultural, human and not, are not bifurcated in these oscillations but attempt to be entangled otherwise. This movement is not understandable along the reductive lines that oppose romantic environmentalism to a pragmatic noninnocent acknowledgment that there is no such thing as "nature" (Morton 2009). Couldn't holding these apparently contradictory positions together open unexpected avenues to think environmentalism? Ecofeminist work can be an inspiration for keeping thinking within these uneasy tensions, as it attempts holding together the feminist aspiration "to explain and overcome women's association with the natural" and the ways in which ecology attempts to "re-embed humanity in its natural framework" (Mellor 1997, 180). In any case, I have never encountered in my dealings with this movement a pure longing to an idealized natural human being who would find natural redemption through ecological immersion. There is a fair amount of awareness among permaculturists about the technoscientific context, about this human practice being a trial-and-error effort of imperfect beings attempting to fray more flourishing ways into ecological futures, acknowledging that we are as much earthy creatures as implicated inheritors of the patently poor environmental record of human history. For the better and worse, this is an alternative tradition that emerged among the technoscientific offspring of the industrialized Global North, the West (something that the charge of whiteness and privileged background of permaculturists often highlights, prompting to challenge and extend its constituencies).[4]

It is, however, at the crossing of these two angles of discussion—everyday ecological ethics and posthumanist naturecultural involvement—that I reach the edges of (my) understanding of care as an ethics, intensifying the speculative tension. Permaculture invites us to think with the "edges"—of lands and systems, where the encounters are both challenging and diversifying beyond the expected and manageable. So there we go. Embedded in the interdependency of all forms of life—humans and their technologies, animals, plants, microorganisms, elemental resources such as air and water, as well as the soil we feed on—permaculture ethics is an attempt to decenter human ethical subjectivity by not considering humans as masters or even as protectors of but as participants in the web of Earth's living beings. And yet, or actually, *correlatively*, in spite of this nonhuman-centered stance, of the affirmation that humans are not separated from natural worlds, permaculture ethics cultivate specific ethical obligations for humans. Collective-personal actions are also moved by ethical commitment and an exigency to respond in *this* world. Possibly, this ethically decentered form of obligation conveys a tension but not, I believe, a contradiction. Asking questions about naturecultural ethical engagements of "nature working" agents brings more fraught questions around the obligation of care in more than human worlds: What notion of ethics is at work in principled stances that aim to decenter humans' position on Earth while still stating its specific obligations? Surely ethics cannot be decentered in this way if it remains attached to presiding over the moral actions of rational, individual, obviously human subjects. But then why would something like ethical "agency" be needed if it is distributed in "nature working" more than human agencies? Isn't this plain anthropomorphism? The question I asked at the beginning of the book remains: Rather than diluting obligation as we eschew human-centered ethics, can we redeploy it? In any case, affirming care, as a dynamic and complex way of sustaining naturecultures, requires asking these questions. It requires displaced speculative moves that decenter "ethicality" and place it as a distributed force across the multiple agencies that make more than human relations.

This chapter unfolds through these questions, through reading permaculture care ethics as an example of an alternative path in the politics of
living with care in more than human worlds. I call this an *alterbiopolitics*
to indicate that I'm in dialogue around the meanings of ethical engagement in a politics of *bios*. The first part of the chapter engages with meanings of biopolitics as an approach that brings back everyday maintenance
of life at the heart of a landscape of ethical questioning in which a decentered care ethics could make a difference. Remaining with feminist nonidealized, innocent visions of care ethics, I needed, however, to engage
with a hegemonic sense in which ethics equals the aspiration to a higher
morality or is depoliticized. Refusing to abandon ethics to its recuperation
and rather engaging with its reclamation might intensify vulnerability to
incorporation into the hegemonic, but also, I hope, the capacity to engage
with the possibilities of displacing contemporary biopolitics' reduction
to the preservation of human life. The second part of the chapter comes
back to reading the speculative possibility of permaculture ethics through
feminist approaches. I emphasize here both the everyday significance of
personal-collective ethos transformation as well as the formation of ethical obligations in naturecultural ecosmologies where human bios is in-
extricable from other than human existences.

Ethics Hegemonic

When engaging with ethical practices at the level of everyday living, it is
difficult to ignore that we live in the "age of ethics." And in transition from
the first part of this book, the politics of knowledge seems to be a good
example to expose how this form of ethical hegemonic thinking can be at
work. The production of knowledge within institutions shows the inflation
of an ethics fully incorporated in the knowledge economy. ELSA (Ethical,
Legal & Social Aspects) is a policy embedded in most Western govern-
ments' policy on science and technology as an institutionalized requisite for
any public funding of research, and numerous research programs and strategic areas favor the inclusion of an ethical "work-package." The European
Union (EU) research funding has a specific research subarea on ELSA for
the programs of Life Science and Technologies. But more generally, all
strategic areas defined by research funds such as "Science and Society" and

the like include ethics as a major area that should be addressed in every research project.[5]

In chapter 1, I noted that the social studies of science have contributed to superseding the traditional view that politics are external notions to the actual practice of basic science—that is, that social, cultural, and political issues only influence science after the hardware moment of technological development or are only related to the uses made of science and technology once "in society." It could be argued that so it goes with ethics. The institutionalization of ethics could just confirm a socialization of science. More and more today, the ethics of research is not uniquely considered the task of ethicists, but social scientists and humanities scholars are required to fill in the "ethicreal" part of the grant application as part of tasks of an "integrated" social scientist (a role that comes with its own modes of managing authorized modes of caring; see Viseu 2015). This is an implicit recognition that "the ethics" is not an isolated moral struggle in the head of a scientist deciding between good and bad action. This presence of ethics in scientific production can take very different forms. On the one hand, it remains vague, for example: such and such scientific issue has "ethical implications," acknowledging that "ethical factors" shape the acceptance and development of science and technology together with "political concerns," "cultural values," or "institutional contexts." This is also the case with references to ethics common in the social sciences in general, showing not so much a proliferation of comprehensive ethical theories or programs but a generalized reference to the *relevance* of ethics that has spread outside specialized realms such as applied ethics, bioethics, and scientific research ethics and well beyond the discipline of philosophy or ethical regulation. So while ethics as a vague form of "self-reflection" is especially notable in the social sciences, it is more generally that a sense that everything has an "ethical" dimension has installed itself in scientific and academic contexts.

On the other hand, in sharp contrast with this elusive omnipresence of ethics, we can observe a highly normative, also all-encompassing, "risk management" approach to "the ethics" of research in everyday legitimation strategies at work in organizations and institutions dedicated to producing knowledge. In the social sciences, a formalized regulation of research procedures often translates into a "tick box" approach, in which "ethics" becomes programmatic and formulaic—another accountability apparatus (Boden, Epstein, and Latimer 2009). From both of these perspectives—a vaguely moralized domain of research and an empty regulatory framework—"ethics" has become an overarching order that traverses all disciplines. The hegemony of ethics can be seen, at least partly, as an inheritance of commitments, however flawed, toward a more just and livable world. Yet in many circumstances these commitments substitute institutionalized ethics for social and political justice protecting the "vulnerable." We can think, for instance, in the context of transnational drug development, of the always improbable notion of informed "consent" by the colonized subjects of clinical trials (Sunder Rajan 2007; Petryna 2009; Dumit 2012). Here ethics becomes a tool for legitimating and maybe paving the "progression" of technoscience and *bioentrepreneurialism* (Latimer 2010b). But hegemonic ethics goes well beyond the domains of academic and scientific risk management. That we live in the "age of ethics" is perceivable in an inflationist use of the word: from corporate ethics to everyday living—garbage recycling, fair trade—every sphere of practice seems today to cultivate ethical awareness as well as produce its own set of ethical codes or recommendations. Such processes have been ongoing for some time now and increasing. In 2007, ethics was reported to be the largest field in growth in philosophy departments in the United States (Bourg 2007). And so engagements with the ethical have exceeded the specialist realm of philosophy. In turn, hegemonic ethics has not escaped the radar of critical thinking across many disciplinary realms, from bioethics (Shildrick and Mykitiuk 2005; Stuart and Holmes 2009; Wolfe 2010) to business ethics (Jones, Parker, and Ten Bos 2005). Questions are asked here as to whether ethics, as it is performed in different sites, reinforces rather than challenges established orders.

In what follows, I capitalize "Ethics" to refer to these modes of ethical normalization, to ethics hegemonic and Incorporated, in contrast with the vibrancies of anormative or not yet normative ethicalities I am trying to engage with. One could also consider abandoning the notion for the good reasons that its transformative potential has been diluted. Indeed, there is nothing groundbreaking in claiming attention to ethics per se when "everything is ethical." I continue to engage with possible meanings for ethics to remain in the impure business of working for a difference within worlds that we would rather not endorse but to which we are not immune—and as I affirmed earlier, because through impure entanglements rather than enlightened distance a critical vision can hope to connect and produce a relevant intervention.

Awareness of the colonizing uses of Ethics and the particular forms of biosocialities that are produced in these processes is important for a decentered reclamation of ethical aspiration in technoscience and naturecultures. In particular, as I mentioned above, engaging with ethics in the context of Ethics hegemonic exposes depolitized engagements with ethics either by diluting them in vague moralizations or by turning them highly normative, though fairly empty, orders of compliance. Depolitization has specific import for ecological ethics, where the impact of individual agency (e.g., "ecological living") is often deemed insufficient with regards to the collective grand-scale policies and radical eco-social transformations needed to confront contemporary environmental challenges (such as climate change). In this context, politics and ethics seem bound to be discussed together: whether it is to oppose, contrast, or correlate them—for example, the ethical, personal, irrelevant option of taking shorter showers versus the significant political option of shutting down all the coal stations (Jensen 2009).

And indeed, everything becoming "ethical" has different origins and implications than the not-so-outdated but more radical process leading to affirming that "everything is political." Critical thinkers have good reasons to look suspiciously at the "ethical turn." When political problems are reduced to ethics, they tend to become individualized, contained in the domain of personal "choice" or lifestyle, seemingly depoliticized as custom or culture or reduced to the ensuring of minimal humanitarian subsistence. Reinforcing this is the sense in which to affirm ethical commitment seems nowadays more acceptable, neutral, and less confrontational than to affirm political commitment. The *ethicization* of the political seems to reduce it to the private domain, personal everydayness, as marks of desertion of political collective transformation (for a counterargument, see Bourg 2007). This mode of prevalence of ethics does seem to confirm a further *depoliticization* of social life in neoliberalism. This seems to make even more sense from the perspective of classic theories for which the ethical usually belongs to the level of individual morality. Following the Aristotelian line taken up by Hannah Arendt, the ethical/moral realm belongs to the private dealings of a person, particularly to the way her own "self" lives in accordance with the good. Ethics refers to a distinct set of negotiations from those happening in the political domain understood as the "public" of the polis aimed at collective intervention—even if, as contemporary virtue ethicists have insisted, in classical theory, "practical wisdom" in the search of "the good life" is developed also by life in the public sphere (for further analysis, see Collier and Lakoff 2005, 26). In other words, ethics is a personal affair but one that is only noble insofar as it aspires to leave a mark in a collective—that is, a polis.

But what does the public/private division that grounds these distinctions mean, when everyday ordinary maintenance of "life" has become so central to the political, seemingly challenging traditional hierarchic relation of the public and private? It is not only an autocritique of feminist politics as assimilated by neoliberalism that is at stake. Keeping within a classic perspective on ethics as a realm of personal edification, this is a particularly bad move: if *even* the political comes to conflate with the private, the personal and ordinary, and worse, with the biological continuation of life, the ethical building of the person cannot but further withdraw her from the more noble affairs of the polis. Individuals are further distanced from ethical life—a process of moral edification of a higher self—as they *descend* into the minor petty matters of maintaining everydayness. Ethical life is even more diverted from the social production of "being," ascribing humanity to the biological constraints of reproducing. From signifying a distinctive and greater form of life for social and moral beings, human existence is reduced to the generic substance of corporeal life and its biological continuation—what I refer to here as *bios,* by contrast with more metaphysical notions of life. In all possible ways, this hierarchical conception of degrees of ethical value of human agency as lesser than its social and political undertakings confirms the long-standing denigration of ordinary living and ordinary care to matters of subsistence rather than existence: the "mere" continuation of "natural" biological life keeps us far from our edification as social and moral beings—a historic denigration that has been a privileged target of feminist critics. As we will see, this hierarchical bifurcation of processes of mattering is also of special importance to an ethics that aims to embrace more than human agencies in the webbing of care. So while diagnoses of a reduction of politics to ethics or a *depolitization* of ethics in the age of Ethics hegemonic seem to go in different directions, both associate ethical and political degeneration to a fall in the domain of the "personal" and "private."

And yet I do not want to dismiss concerns about how the "age of ethics" dilutes the significance of ethical as well as collective political action. In particular, as we will see, it is worrisome that diluting ethical agency in an amorphous way, "everything is ethical" can lead to rendering obligation and commitment indistinguishable from agency tout court. In other words, if every personal action is an ethical action, ethical *commitment* or response-ability makes no particular difference—nor does the building of oppositional collectives. And yet, precisely because of these concerns, I want to take seriously the significance and potential of the implosion of politics with the ethics of everyday practices dedicated to the everyday continuation of life. There are indubitably many reasons, and ways in which, to criticize how technoscience works today with Ethics. Yet just debunking Ethics, or a blanket rejection of the spreading of ethics as depoliticization, not only would be one of those gestures of distant critique that I'm trying to avoid, it would obscure possibilities emerging in terrains where the meanings of ethics are being reconfigured.

This more hopeful outlook for the possible politics of ethical engagement is driven by a recalcitrant attachment to prolong feminist affirmations that "the personal is political," resisting to abandon this insight's collective inheritance to its assimilation and recuperation by a moral order that privatizes-personalizes politics. Also, at the very moment where the political significance of the maintenance of *bios* through all Earth beings' everydayness is dramatically exposed, regretting the emphasis on "personal" everyday agency as sociopolitical decay can only but confirm the historical ethico-political disengagement with the "life domain"—reduced to "biological" life and devalued with regard to the higher realms of social beings. And, as mentioned above, we can see this in how the notion of human "reproduction" of life remains tainted and the fleshy biological processes shared with other living beings denied, with regard to production that remains the higher level of human edification—a bifurcation of nature that even the moves to revalue as "social" the contribution of reproduction tend to ratify. While we could continue pondering whether human ethical agency should be sociopolitical rather than "individual" or merely biological, not only do we confirm the classic binary, but the everyday corporeal life of everything in this planet continues enduring pervasive technoscientific intervention in the very matter of biological existence, affecting the integrity of all beings on this planet with wide ecological disruption. In what follows, I engage with biopolitics as a mainstream ethical discussion that is open to the significance of practices at the level of everyday *bios*, recognizing forms of ethical agency that do not correspond to social/biological, political/ethical, collective/individual bifurcations. These ethical discussions offer, as we will see, an interesting contrast for thinking the singular combination of a personal/collective ethics characteristic of permaculture.

Our Bios, Our Selves

Technoscience doesn't study biophysical actualities but (re)makes them and commodifies them, coproducing new forms of worldwide relationality and living (im)possibility. The pervasiveness of technoscience in the living world raises a justified sense of urgency to further embed ethical engagement at the level of *bios*—including tackling with the economic pressures to extract "biocapital" (Sunder Rajan 2006; Cooper 2008) from human "biological labor" (Vora 2009b). But while reference to ethics becomes more pressing in contexts dealing with technoscientific biopower, it is also here that the limits of classic ethical theory and institutional bioethics have become more salient.

Nikolas Rose's articulation of a "somatic ethics" for "biological citizens" marked an intervention merging the ethical and the political in the domain of bios (Rose 2007). This approach is relevant for me here because it acknowledged that we live in an "etho-political" age in which political issues have become problematized in terms of ethics. And this "ethicalization of politics" is particularly visible in the worlds where politics *are* biopolitics and in which "value-driven debates" follow bioscientific development. It is in such a context that bioethics has become a "necessary supplement" for the public acceptance of decision-making (97), tending to represent institutional regulatory frameworks that legitimize, amend, or pave the way for biotechnological transformation. Rose displaced bioethics with an idea of somatic ethics (from *soma*, the body) to designate a form of bioethical engagement emerging from communities coping with the politics of their "corporeal" existence (257). Somatic bioethics recognized that biopolitics happens in people's concrete embodied everyday practices and not only in institutions, ethical committees, or even citizen groups. This approach indeed relocated ethics at the level of ordinary living and initiated two interesting displacements.

First, the "bios" of biopolitics is quite different from the general idea of (social) life engaged by those concerned by forms of power aimed at controlling people's existence at every level of experience and subjectivity—as well as to the forces that confront or escape this power by producing alternative subjectivities and forms of collective living (Papadopoulos, Stephenson, and Tsianos 2008a; Hardt and Negri 2009). These approaches offer continuation of debates around a Foucauldian vision of "biopower" understood as the normalization of life through the control of human populations and selves. However, as Donna Haraway pointed out, Foucault's biopolitics were a "flaccid premonition" (Haraway 1991a, 150) of what contemporary technoscience implies for everyday *bios*. A general idea of (social) life does not grasp the transformative character of technoscience that intervenes at molecular and genetic levels and has significant effects on the wider planetary ecosystem. Contemporary biopolitical ecosmologies—across a number of fields of practice in STS, New Materialisms, Environmental Humanities—recognize a world where power not only works through social normalizing but acts with and from biology, organisms, cells, genetic makeup—a "politics of matter" (Papadopoulos 2014b).

Second, ethics as a notion is also displaced. Ethical agency in perspectives such as "somatic ethics" concentrates in human life as affected primarily by (biomedical) technoscience. Rather than focusing on how biopolitics affects the "ontological" status of the "human" (Agamben 1998), this requires considering ethical disruptions in specific corporeal ways. Like other forms of critical ethics, the stake here is in diverting from universalizing conceptions of the ethical subject as an autonomous, rational, and defined "self"

(Stuart and Holmes 2009) to focus on the ethical as it is affecting bodies in processes of change (Shildrick and Mykitiuk 2005; Heyes 2007). Though still human-centered, these ethics here are not about individual rationalizations or about a normative identification between the rational and the good. These ethics are better understood as developing within what Collier and Lakoff call a "regime of living": "situated configurations of normative, technical and political elements that are brought into alignment in problematic or uncertain situations." These involve forms of living that have a "provisional consistency or coherence" but not really the "stability and coherence of a political regime" (Collier and Lakoff 2005, 31–33). Such collective arrangements are not primarily founded on an individual as arbiter with standards of judgment of what is properly or improperly moral. In other words, the ethical of biopolitics in technoscience is not about stable norms of morality managed among humans; it includes a range of elements, sociotechnical forces, and practices and doings of agencies constantly reconfigured in function of *material* conditions in specific situations.

These engagements with biopolitical regimes of living open paths for a speculative ethics in more than human worlds along two displacements: engaging ordinary personal practices *as collective* and pushing toward a decentering of ethical subjectivity. They also support the search for nonnormative approaches to ethics. Ethico-socio-technical everyday assemblages are approached at the level of the unexceptional everyday, they are objects of sociological or anthropological study in a very different way than it was for classic moral theory. The "ethical" attracts the attention of the biopolitical social scientist as an important element to understand the emergence of new social forms rather than for promoting a particular ethical (or political) obligation according to a "normative" stance on moral subjects facing grand Ethical dilemmas. This more processual approach affects the way ethical agents are envisioned in new sociologies/anthropologies of ethics in biotechnologies. Individuals are not at the source of rational decision-making regarding biomedical choices. Nor are collectives clusters of individuals managing the mastery of their agency. All are embedded in the biopolitical fabric in fairly unpredictable and emergent processes. Bodies (soma) or situations (regimes) are seen as sites where sociopolitical interests and scientific developments touching "life itself" coalesce.[6]

But doesn't such a move to decenter subjects and their moral norm dilute the possibility of ethical obligation? Not necessarily. Even decentered from the focus on an individual rational subject, some form of ethical subjectivity remains crucial in the forms of biosociality that have focused the attention of most contemporary biopolitical engagements. Actually, the approach to ethical obligation remains fairly traditional in that the challenges are mostly attached to a human's biological life considered in terms of people dealing with their corporeal existence, with their body-self, or their "environment." Our bios, our selves. Conceptions of biopolitical ethics such as somatic ethics start from an obligation to care for one's own body, personhood, and, by extension, that of proximal ones or a community gathered around a biomedical issue (e.g., patient groups). This is understandable given the types of collectives that prompted the exploration of biosocialites—such as individuals creating collectives (citizen and patient groups) out of concerns with their bodies, their relatives, or the future of kin. More is needed to disrupt today's overwhelming focus on the privatization of responsibility and the moral pressures to take ownership of our biological destinies. They can also confirm the hegemony of "self-care"—and, by extension, of our "dependents"—often corresponding to a de-responsibilization from the shared burdens of collective health care and welfare (Stuart 2007). So while the focus of ethical action is firmly placed on matters of ordinary maintenance of corporal life, the identification of ethics with matters pertaining to the "private" life of humans as individuals remains unchallenged.

In order to shift the perspective on what counts as an ethical intervention in biopolitics, in order to understand the committed difference in the hegemony of diluted ethics that is being made by the personal-collective ecological practices of movements such as permaculture, we need two additional speculative moves. First, to interrupt even further the association of "personal" ethical engagement with the "individual" and the "private." Coming back to thinking with the feminist insight that "the personal is political," personal ethico-political practices of change need to be also rethought as collective. How else could we pay attention to situations when people change their ways of doing at the level of personal everyday life but would not think of this as an individual or private action or even

consider doing it if outside a collective? My questioning remains inspired
by feminist visions of care ethics built from standpoints grounded in collective understandings of women's work of maintenance of everyday relationships. These approaches saw the troubles of personal labors of everyday
care as part of a larger societal disengagement from their importance. Personal work to transform the ways society deals with caring for others in
the everyday was brought upon by a collective rethinking made possible
by women's movements. Such a way of engaging with the problem of care
as something that can be done individually, but is always interconnected
within collective endeavors, is very different from care that starts or aims
at self-care—it is also different from advocating pastoral care of the state
for its subjects and from a Foucauldian-inspired "care of the self" (Foucault 1990, 1988). I will come back to this aspect, for now I want to emphasize that in order for reclamations of the political significance of everyday
"personal" experience (Stephenson and Papadopoulos 2006) not to simply
ratify the hegemony of diluted ethics ("everything is ethical") and notions
of self-care, we need a notion of everyday ethics as agency that is invested
by collective commitments and attachments. The point is not to dismiss
the political importance of biosocialities but to argue for a displacement
of ethopolitics in biopolitics that brings us closer to challenge what we
include in bios as a collective in search of as well as possible relationalities.
This is because interrupting the identification of ethical agencies of biopolitics to concern and care for the preservation of one's individual body/
self—or, by extension, that of one's kin, children, family, fellow citizens—
requires disturbing a vision that conceives of human survival and well-
being independently from the rest of Earth's beings and thinking care on
the grounds of nonanthropocentric, naturecultural ecosmologies.

Naturecultures—Decentering Ethics

Naturecultural thinking is an ecosmology of affirmative blurred boundaries between the technological and the organic as well as the animal and
the human—whether this is considered to be a historical phenomenon,
an ontological shift, and/or a political intervention. Naturecultural thinking has been at work in the humanities and the social sciences, together
with relational ontologies that engage with the material world less from

the perspective of defined "objects" and "subjects" but as composed of knots of relations involving humans, nonhumans, and physical entanglements of matter and meaning (Barad 2007). Naturecultural thought is also invoked to name a strand of thought in the social studies of science and technology. As we saw in chapter 1, radical constructivist approaches in this field—actor-network theory, in particular—questioned the existence of such thing as "the social" to bring attention to concrete practices of world-making in which agency is distributed between actors that are not only human (or to include objects as agents in the production of sociability). Attention is drawn to the agential significance of entities that go from the microchip to the molecule, from the robot to the primate and the microbe. Naturecultural visions in this context also challenge epistemic bifurcations of nature and share with sociotechnical imaginaries a shift of attention to nonhuman ways of life and an awareness of the ontological connectedness between multiple agencies and entities. They "dis-objectify" nonhuman worlds by exposing their liveliness and agency; they "de-subjectify" the human by trying to think of it as a form of ontological agency among others. As such, they promote a mode of attention that resists falling automatically into the "human" perspective.

I recall these general trends to note their common potential to contribute a conception of ethics that decenters the human subject in biopolitical collectives in technoscience. Social studies of science can be particularly helpful to approach ethics within complex and emergent fields (Ong and Collier 2005), observable as actor-networks, as becoming visible through novel entanglements, attachments, and detachments (Palli Monguillod 2004; Latimer and Puig de la Bellacasa 2013). Paying attention to ethicality in practices, in entanglements of relationality and distributed agency on the ground, is a way to research ethicality attuned with an attention to specificity that refuses to start thought from a normative perspective. These materialist ontologies have the potential to displace ethical research beyond its focus on moral orders and human individual intentionality. They enrich our perception of complex articulations of agency that involve associations between humans, nonhumans, and objects working in the realization of new relational formations. They could then contribute to a "postconventional" (Shildrick and Mykitiuk 2005) vision of the ethical that embeds it in processes, rather than discussing it as a set of added concerns that humans reflect on when technoscientific and other material matters are already established.

Of course the broad field of STS has not remained immune to the "age of ethics" discussed earlier: references to the ethical have become more and more frequent—in combination with, or replacing, earlier concerns for elucidating the political interests supporting science and technology. However, as in many of the approaches to biopolitics considered above, the ethical remains in this field of study an ethnographic or sociological object. A general perception has remained that STS scholars avoid taking explicit *judgments* or elaborate *prescriptive* frameworks: "their job is to illuminate the social processes by which arguments achieve legitimacy rather than to use their understanding of those processes to establish the legitimacy of their own arguments or positions" (Johnson and Wetmore 2008). Like with the Latourian approach examined in chapter 1, interest in the ethical in this sense is not so much aimed at fostering ethical obligations or affirming commitments but remains mostly about observing ethical issues under construction within sociotechnological problems and detecting the participants "assembled" in this making. Thus, in spite of the potential of approaches in STS to transform the ethical, it is rare to see its insights thematized as possibilities for proposing new ethical visions.

From an ethicist's perspective this could be seen as a normative "deficit" (Keulartz et al. 2004). However, to identify ethical engagement to normative claims is a reductive approach that allows overlooking other potential contributions. As I noted in the introduction, I have followed Suchman's cue to STS scholars when she reminds us that "the price in recognizing the agency of artefacts need not be the denial of our own" (Suchman 2007b, 285). And Karen Barad's asubjective approach to ethical agency offers a most prominent attempt to engage ontologically with the ethicality of matter (Barad 2007). Following Suchman and Barad, I keep attempting to open speculative paths for a notion of care that responds to ethical commitments and obligations at the heart of ideas of distributed ethical agency: if the ethical is complex and emerging, this also involves chances to contribute to its shaping. Engagements to approach the ethical as an "upstream" approach to the creation of technological innovation in the form of interventions could also make a difference in fostering an "alter-ontology" (Papadopoulos 2011) rather than confirming existing ontologies by following and describing its operations. Interventions in co-shaping do not necessarily need to be a normative move by which an "enlightened" social scientist or humanities scholar would put on the ethicist hat and adopt the role of an arbiter pointing out the right and wrong ways to go in the technoscientific moral maze—but as an immersed participant in the field. More than following the actors, less than showing "the" way.

Nevertheless, the disengagement with ethical theorizing (and position-ing) in STS not only responds to a distance from normative perspectives. What makes it more compelling is its consistency with a rejection of the humanist frameworks—in which ethics is traditionally understood. Natu-recultural cosmologies require a form of ethical commitment attuned to this decentering of human agency. One has to note, however, that the category "nonhuman" in studies dealing with science and technology, though a helpful one, also tends to conflate very diverse forms of life. This is important because decentering human agency will have different impli-cations whether we refer to engagements invested in the dis-objectification of the "natural" (modes of life in *bios* and *phusis*) rather than of the "tech-nological" (*techne*).[7] Not only does each human–nonhuman configuration have its specificities, but the interference of the "nonhuman" in the ethical and the political varies generically whether attention is turned to artifacts or animal/organic entities. This is not only a conceptual issue or a matter of ontological categorization; it is a concrete problem. If we aim to think the ethical not as an abstract sphere but as embedded in actual practices, when dealing with assemblages that involve organic and animal entities we enter a world populated by particular worries around, for instance, animal rights, domestication, ecological movements, resource exhaustion. Also, we touch affective spheres associated with living bodies such as suf-fering. So though in naturecultures it is pointless to *separate* the entangled worlds of *bios* and *techne*, it seems also vital to recognize ethico-political specificities especially when "nonhuman" involves engaging with alterities that are capable of being affected by human intervention with pain, death, and even extinction (Bird Rose and Van Dooren 2011; Van Dooren 2014), as well as to respond by creating affective and life-sustaining interdependencies (Haraway 2007). Acknowledging agency and liveliness of sentient beings is not the same as recognizing that machines are "alive." Also, the material semantics of naturecultures when they concern *bios* might then be less those of networks and connections than those of ecologies and relations (Puig de la Bellacasa 2016). The engagement with nonhuman others from the animal/organic world produces a different set of ethical concerns than the engagement with technological entities. Things are not one thing—like humans are not "the" human (Papadopoulos 2010).

In naturecultural ecosmologies, agency is distributed and decentered from its humanistic pole. But here the ethical consequences of interdependent entanglements of nonhumans and humans are not only about the preservation of human existence and/or about which decisions will better respond to novel forms of biopower introduced by technoscience—for example, the effects of biomedicine for human subjectivity, of technological waste on human health and their environments. Other problems have become particularly visible for these interventions thanks to critical animal studies and the environmental humanities as well as, of course, ecofeminism, animal right movements, and indigenous struggles: How do we actively engage with the lived experiences of forms of nonhuman *bios* whose existences are today increasingly incorporated in the cultural world of human *techne*? How do we acknowledge "their" agency, and our involvement with it, without denying the asymmetrical power historically developed by human agencies in *bios*? How do we engage with accountable forms of ethico-political caring that respond to alterity without nurturing purist separations between humans and nonhumans? How do we engage with the care of Earth and its beings without idealizing nature nor diminishing human response-ability by seeing it as either inevitably destructive or mere paternalistic stewardship?

The sites abound for exploring situated pragmatic ways of addressing these questions in creative ways (e.g., animal carers, conservation practices). Hoping to contribute to these efforts, and having set out a perspective on the field of tensions around more than human ethicalities, I am coming back to my own experience with permaculture collectives to propose a reading of this movement's ethics as an alterbiopolitical intervention in naturecultures that builds ethico-political obligation on personal-collective practice in a more than human decentered way.

Permaculture as Ethical Doings

Permaculture practices are ethical doings that engage with ordinary personal living and subsistence as part of a collective effort that includes nonhumans. They decenter human agency without denying its specificity. They promote ethical obligations that do not start from, nor aim at, moral norms but are articulated as existential and concrete necessities. These ethics are born out of material constraints and situated relationalities in the making with other people, living beings, and earth's "resources." Thus, the "principles": care for the earth and people and return of the surplus are both quite generic—their actualizations vary—and involve design principles, that is, very concrete, specific, material, and sometimes inescapable ways to work with patterns of bios (ecological cycles, physical forces). Among people who have followed these trainings, stories abound about their subsequent attempts to implement the practices they learned—in local communities both in urban and rural environments, from a backyard to the local council, or joining larger ways of public eco-activism. Many strongly insist that the trainings and other collective ways of engaging in permaculture have changed their personal everyday ways of relating with food, plants, animals, technologies, and resources, and affected how they valued their own impact on the planet in smaller and bigger ways. Activities can go from starting to compost food waste, to plant and produce food locally, to promote ecological building. But even when actions are acknowledged as deeply intimate or individual—as can be a spiritual connection to a tree, or the building of one's self as a more ecological person—they are affirmed as collective.

The "collective" here does not only include humans but the plants we cultivate, the animals we raise and eat (or rather not), and Earth's energetic resources: air, water. It is in connection with these that human and nonhuman "individuals" live and act. At every level of human subsistence we depend on them—and in these specific contexts of eco-design painfully aware of ecological disruption—*they* are considered as also depending on us. And as such, humans exist only in a web of living co-vulnerabilities.

Permaculture ethics of care are based on the perception that we are embedded in a web of complex relationships in which personal actions have consequences for more than ourselves and our kin. And, conversely, these collective connections transform "our" personal life. The ecological perception of being part of the earth, a part that does its specific share of care, requires Earth not to be a spiritual or visionary image—for example, Gaia—but is felt. Earth as "real dirt under our fingernails" (Starhawk 2004, 6), and that our bodies are conceived materially as part of it, for example, responding to the needs of water because we *are* water (Lohan 2008); human energy, including activist energy (Shiva 2008), being a living material processed by other forms of life. So while permaculture ethical principles can indeed be read as ideas that practitioners become able of transforming into doings, I believe it is more accurate to say that it is the ongoing engagement with personal-collective doings that gradually transforms the way we feel, think, and engage, with principles and ideas. Ongoing doings thicken the meanings of the principles by, for instance, requiring that we learn more in order to know the needs of the soils we take for granted (Ingham 1999) or other biological and ecological processes, such as the water cycles.

Before continuing, I want to mention a simple example of one of such ethical doings: practicing composting. For people living in urban areas composting is a more or less accessible practical technique of caring for the earth, an everyday task of returning the surplus and aiming to produce "no waste" (Carlsson 2008, 9). It is a relational practice that engages ways of knowing. A good compost is not just a pile of organic waste, and therefore compost techniques are an important part of Earth activist trainings. Not only how to keep a good compost going, but also how to become knowledgeable regarding the liveliness, and needs, of a pile of compost. For instance, one can check if a pile of compost is healthy by attending to the population of pink sticky worms. Worms, in compost—some people keep worm buckets in their kitchens—are a good example of the non-human beings we live with and of which permaculture ethics makes you aware, but not the only one: "anyone who eats should care about the microorganisms in the soil" (Starhawk 2004, 8). Here naturecultural interdependency is not only more than a moral principle, it is also more than a matter of fact—or technique—that we become aware of. It becomes a matter of care to be involved in through ethical doings.

I am interested in how this "we should care" doesn't work without a transformation of ethos by which obligation emerges within a necessary doing, as well as doings that transform or confirm obligation. I emphasize the word "doing" to mark the ordinariness, the uneventful connotation of this process, in contrast with "action" or distinct moments of decision-making or other ways of delineating ethical events. Obligation toward worms is a good example of doing-obligation. Worms are a more visible manifestation of soil life than microorganisms, but they are as easy to neglect. Caring for the worms is not a given: most people have learned to be disgusted by them. In permaculture trainings they become a signifier of a transformation in feelings as we are invited to appreciate them—"worms are the great creators of fertility. They tunnel into the soil, turning and aerating it. They eat soil particles and rotting food, passing them through their gut and turning them into worm castings, an extremely valuable form of fertilizer, high in nitrogen, minerals and trace elements" (Starhawk 2004, 170). Becoming able of a caring obligation toward worms as our earthy companions in this messy and muddy way is nurtured by hands on dirt, curiosity, and even love for the needs of an "other," whether this is the people we live with, the animals we care for, or the soil we plant in. It is by working with them, by feeding them and gathering their castings as food for plants, that a relationship is created that acknowledges these interdependencies: while some still might find them disgusting, this is not incompatible with a sense that these neglectible sticky beings appear as quite amazing as well as indispensable—they *take care* of our waste, they process it so that it becomes food again.[8]

This caring obligation is not reducible to "feel good" or "nice feelings"; repulsion is not incompatible with affectionate care (as anybody who has ever changed a baby's soiled diaper or cleaned up the vomit of a sick friend might know). Neither is this obligation to care for an interdependent earthy other understandable as a utilitarian one—I take care of Earth, via soil and the worms, because I *need* them, because they are of use to me. It is true that some of the teachings of permaculture techniques emphasize that when we don't listen to what nonhumans are saying, experiencing, needing, the responses are consequential for us, too—from the everyday failures and mistakes faced by every grower, to the extinctions and animal-related epidemics among many other failures of care. In contrast, other Earth beings are not discussed as existing to serve "us"—on the contrary, utilitarian approaches are constantly challenged, and the notion that nature provides "services" (see chapter 5) is not characteristic of permaculture. But if this is not a utilitarian relation, it is not either an altruistic, self-sacrificing one, where nature has value for "itself." While this traditional debate on altruistic versus utilitarian environmental stewardship might be important in other settings (for a discussion of these debates, see Thompson 1995), here it precludes a speculative engagement with what could be becoming possible in this specific conception of relationships and mutual obligation where living-with rather than living-*on* or living-*for* are at stake.

As I mentioned earlier, human agency in the permaculture ecosmology is nature working. This means that humans are full participants to the becoming of natural worlds. However, they have their own worldly tasks—their own naturecultural ways of being in this relation. Creating "abundance" by working with nature is affirmed as a typical human skill and contribution. Yet abundance is not considered a surplus of life (as yield) that can be squandered, or as self-regenerative biocapital to invest in a speculative future (Cooper 2008). On the contrary, it is only by returning the surplus of life—for example, by composting—that the production of abundance can be nurtured. Working-with-nature is something that permaculture activists consider wisdom shared and maintained by alternative agricultural practices that have somehow survived within or in spite of industrialized agriculture, for instance, in the syncretic practices of contemporary indigenous populations. Starhawk cites Mabel McKay, a Powo healer: "When people don't use the plants, they get scarce. You must use them so they will come up again. All plants are like that. If they're not gathered from, or talked to and cared about, they'll die" (quoted in Starhawk 2004, 9; see also Mendum 2009). And yet, that human workings in the ecologies we engage with are vital doesn't mean they are at the center. The irony is that it is considered a typical aim of good permaculture to be able to reduce human work as much as possible. In some places, the role of human agency might be to let be. For instance, some plants should be ignored because they are not there for humans, but for others—animals.[9] It could be said that letting other than humans be, a kind of conscious neglect, is also part of the task of care. This points to a hesitation that some beings might be out of reach of care for better or worse or might require a form of ethicality that attaints the limits of embodied relational efforts, as in the case of extinct species that we cannot sense (Yusoff 2013). What I am seeking here is not to delineate a universally reaching imperative of care that would define human relations with all Earth beings but to specifically learn from these doings of care that include practical, particular, shifting relations where humans are involved with other than humans in ways not reducible to a human-centered-use and are also radically naturecultural.

Once again, my insistence on this naming attempts to short-circuit the reduction of this ethics to one or the other side of humanist binaries. Of course, one could argue that because permaculturists often present the practice as a better "science" it remains within an epistemocentric humanist vision (Holmgren 2002). But what I have observed in my work with permaculture collectives and permaculture-inspired activism is that humanism and scientism are often advanced somewhat defensively to respond to external identifications of this movement with ecological visions that put "other" beings before humans—for example, considering humans as a separate, destructive, invasive species and science and technology as evil— or that encourage a nostalgic back-to-nature ideal. Beyond this "defensive," or justificatory image (that has nonetheless performative effects in the transformation of permaculture activism in a network of accredited trainings), the accent is rather put on a commitment to the "people" of Earth, inseparably including human and nonhuman beings in a range of different agencies and doings that need each other. Without caring for other beings, we cannot care for humans. Without caring for humans, we cannot care for the ecologies that they live in. Care for "the environment"— as something surrounding "us"—wouldn't be a good way to conceptualize these ethics.

There is another reason why altruist self-erasure or sacrifice (of humans) does not respond better than a utilitarian perspective to these relations. If

I read these practices as marked by a form of biopolitical ethics attuned to naturecultural awareness, it is also because here care for one's body-self is not separable from peoplecare and Earthcare. This movement exemplifies well the interdependence of the "three ecologies"—of self (body and psyche), the collective, and the earth—that Félix Guattari famously called upon with political urgency for the near future, believing that none could be realizable without the other (Guattari 2000). As Starhawk considers, material-spiritual balance cannot be attained through abstract engagement with caring for the earth. On the contrary, the reference to an "ideal" Earth leads "our spiritual, psychic, and physical health" to "become devitalized and deeply unbalanced" (Starhawk 2004, 6). Conversely, in permaculture trainings there is an insistence on not neglecting the needs of one's body-psyche in the profit of "serving"—burnout is taken into account as a typical activist condition. Thus, while activist care of one's self is embedded in obligation toward a collective, it is not considered "healthy," or effective, to ground care in an altruistic ethics in the face of environmental destruction. As Katie Renz puts it, permaculture is "not some last-ditch effort in the emaciated face of scarcity, but a cultivation of an intimate relationship with one's natural surroundings to create abundance for oneself, for human communities, and the earth" (Renz 2003). Admittedly, the aim is not modest, or self-sacrificial. It is not even sustainability. It is abundance. In the same way, the affect cultivated in Earth activist trainings is not despondency in the face of the impossible but joy of acting for possibility. In terms of Joan Haran, here hope is a praxis (Haran 2010).

Ultimately, permaculture ethics is a situated ethics. I am brought back to one of the mottos incessantly repeated in trainings and manuals: "It depends" is the answer to almost every permaculture question. The actualization of principles of caring are always created in an interrelated doing with the needs of a place, a land, a neighborhood, a city, even when a particular action is considered with regard to its extended global connections. Here, "personal" agencies of everyday care are inseparable from their collective ecological significance. It is important to remember that permaculture ethics are not only about planting food or raising animals or sustainable building. In many of its versions, and strongly within the Earth Activist Training tradition, they are also related to public actions of civil disobedience and nonviolent direct action—guerrilla garden creation, public demonstration of techniques in alter-globalization oppositional events, "seedbombing" (Starhawk 2004, 2002; see interview with Olhsen in Carlsson 2008, 74–79). More generally, permaculture ethics are thought also as forms of organizing—for instance, promoting forms of collaborative direct democratic sharing instead of competition. They are not about an abstract external vision of the practices of others but an intrinsic transformation of ethos.

Care—Ethos and Obligation

Until now I have worked with an unexplained assumption: I have thought ethics from the perspective of its closeness to ethos rather than to morality, taking distance from Ethics with a capital E as the enactment of normative stances, a more fixed and vertically experienced domain. Rather than relevant as an Ethics, I have spoken of permaculture principles as ethical doings. And yet this book is permeated by notions of obligation and commitment. Escaping "Ethics" does not mean absence of ethical agency and attention, but it shifts focus to the intensities and gradations of "ethicality" involved in any situation, even, and especially, when Ethics are not (yet) fixed. When Tronto affirmed that care is not reducible to a moral disposition, she signaled the displacement of normative morality by a politics of care. The ways in which we care for the everyday have a quality of "ethicality," embedded in processes of situated relationality, perceivable in ethos rather than in moral attitudes, principles, and discourse.

Thinking this way follows the requirement of looking at the specificity of moments, particular relations, of ecologies where the ethical is both personal agency *and* embedded in the "ethos" of a community of living. Attention to situated specificity is close to a constructivist approach to doings and *undoings* of the ethical embedded in technoscientific assemblages and naturecultures. As we saw previously, ethics can this way become an object of social research, which doesn't see the ethical as an added set of concerns but as entangled in the making of sociomaterial worlds. However, this is not sufficient to consider, among the material constraints embedded in practices, those that across this book I keep calling "obligations." Obligation is a heavy-loaded term in ethical theorizing and moral philosophy, and I must confess that I only became (at least consciously) aware of this well after it had surreptitiously become part of my vocabulary to speak of care. Since then I have been brought to realize that my use of the term runs against much of what it signifies for political theory:[10] justice, contracts, promises, and individual reciprocity. Precisely for these reasons, Tronto had proposed that "a flexible notion of responsibility" was a more attuned concept to a politics of care than obligation—a rather rigid concept in political and moral philosophy (Tronto 1993, 131–32).

My use of obligation originated not in moral theory but in the philosophy of science. It was inspired by Isabelle Stengers's use of the term in her "ecology of practices." Admittedly mine is a rather displaced prolongation of the concepts of Stengers's philosophy of practices, tailored to a project of accounting for the specificity of modern scientific practices (and the historical success of the experimental fact) rather than to think about everyday ethical ethos. However, Stengers's thinking offers paths for this journey, because it avoids both epistemological orders and relativistic accounts of scientific practices. This is relevant for an understanding of ethical obligations of care, which have a contingent necessity that emerges from material and affective constraints rather than moral orders. Speaking of practices of everyday ethos transformation as "ethical doings" is an attempt to avoid defining in advance a code of conduct or a normative definition of right and wrong care. But affirming care as a generic activity doesn't mean that a care ethos is just there, or that its possibility is randomly fostered. This is probably why I find it helpful to think of ethos as marked by obligations, which are, for Stengers, a form of "constraint." Constraints for Stengers "have nothing to do with a limitation, ban or imperative that would come from the outside . . . that would be endured, and everything to do with the creation of values" (42). Constraints are not negative—enforcing—aspects of a practice; on the contrary, they are "enabling" the practice, they make it specific, and develop in close relation to ways of being and of doing. In Stengers's philosophy of practice, constraints are embedded in relations between worlds and (scientific) knowers entangled within a specific setting. Constraints are not "conditions; in that they do not provide an explanation, a foundation or legitimacy to the practice." More important: "A constraint, must be satisfied, but the way it is satisfied remains, by definition, an open question. A constraint must be taken into account, but it does not tell us how it should be taken into account" (43). Practices develop a relational ethos with a world, a process through which material constraints are co-created (Stengers speaks of reciprocal capture). In turn, constraints re-create relational, situated possibilities and impossibilities. Under this category, Stengers then defines "requirements" and "obligations" as a type of "abstract constraints"—abstract in the sense that these become more or less stabilized and can be repeated, transported, translated as the core of a practice and ask to be taken into account for a specific practice to be considered such. Obligations here refer to what obligates practitioners to what is "required of a phenomenon" for it to be addressed as a focus for a particular practice. In this "ecology of practice," constraints, requirements, and obligations hold together a "heterogeneous collective"—competent specialists, devices, arguments, and material at risk—that is, phenomena whose interpretation is at stake. The ways in which these entanglements affirm what is of "value" is an immanent one, therefore comporting a dimension of "nonequivalence"—one relational practice does not equate another (52–53).

While Stengers goes on to develop this conceptual construction for qualifying the event of modern scientific practice, her notions have also a generic quality that pertain to the relation between the immanence of practicing—as it re-creates behaviors and relations in contingent processual manners, an ongoing continuous ethos—and those patterns of behavior that endure and are considered of value. The latter would then include a dimension of translatable and relatively enduring ethicality. And a generic character of these ecological notions is visible in value-creation processes among more than human agencies:

Every living being *might be approached* in terms of the question of the requirements on which not only its survival but also its activity depend. . . . every living being brings into existence obligations that qualify what we refer to as its behavior: not all milieus or all behaviors are equal from the point of view of the living. . . . Viewed in this generic sense, *requirement reflects the normative and risky dimension of dependence* on a milieu, that is, on what may or not may fulfill needs and demands. (55, emphasis added)

The "normative" refers here not to Morality but to those aspects of necessity that define relations of dependency contingent to an ecological milieu and the risks involved in these relations (for instance, when needs and demands are not fulfilled).

In resonance rather than correspondence with this account that betrays the usual meaning of normativity, I read ethical obligations of care as constraints that get to endure across more or less changing relational fields. They transcend specific instances of production of care ethos but not necessarily to become moral norms, or even positions, but because they require engagement with an ongoing doing. These ethical obligations are commitments that stabilize as necessary to maintain or intervene in a particular ethos (agencies and behaviors within an ecology). They are not a priori universal, they do not define a moral, or social, or even natural "nature": they *become* necessary to the maintaining and flourishing of a relation through processes of ongoing relating. The constraints that mark the ethicality in agencies go beyond prefigured conditions—they are not predetermined, but they are neither arbitrary nor random. Relations are always connected to specific worlds; they do specific worlds and create interdependencies in ways that become ethos. From that perspective, where living means entangling ethos and milieu, even moves that may at first appear as individual, strategic, or instrumental have a dimension of affective interdependent entanglement.

The ethicality of practical doings can thus also be envisioned from the perspective of how they generate ways of doing that both endure and change (ethos creation). In caring, an ethos creates its ethics, rather than the other way round. The ethical meaningfulness of practical doings is thus inseparable from constraints, but these are not necessarily moral norms. This is different from explaining ethos as ways of behaving according to preexistent norms and conventions that sort out the good and the bad, the true and the false—or of explaining ethical "choice" as the action of objective self-reliant individuals in a given situation. Morality is neither outside, nor before, nor even after ethos. Rather, it can be said that *norms* and *principles* are particular modes of expression of ethos formation and deformation but do not express the whole of ethical significance. Such are situated explanatory artifices belonging to a historical "mode of thought" prevailing in

(Western) ethical thinking. Thinking ethics from the perspective of its closeness with ethos points at a more immanent conception of how ethics is formed: of the ethical as a social practice, as a living technology with material implications in remaking human and nonhuman ontologies. I see care as one of those doings permeated by ethicality and embedded in a living ethos. It is an obligation that is inseparable from the material continuation of life. This brings ethical obligations of care to a different status in the politics of *bios*; it pertains to modes of maintenance, repair, and continuation of life through ecological practices that unsettle traditional binaries. The ethical obligation to care by which ethos generates commitment happens through engagement with ongoing relative constraints. When caring for and taking care, or having something or someone to care for us, particular actions become *obligatory*: they create and re-create demands and dependencies, they become necessary in a specific world to subsist and thus somehow *oblige* those who inhabit that world. The mutual, albeit multilateral, web of labors of care is fully permeated with ethicality even when agencies are not intentionally ethical.

To engage speculatively with ethicality in the making as nonnormative might require a form of "suspended judgment," of deliberate indecision. But suspended judgment does not necessarily mean ethical or political agnosticism or the dilution of obligation. This is a crucial point for keeping close to the specificities of situated human agencies and response-abilities in more than human webs of care. From the perspective of ethical obligation outlined above, something can be considered good without this consideration being imposed from an outside. This is particularly relevant for commitments and obligations that emerge within everyday practices of mundane "taking care."

We can also take this commitment further, as it resonates with another sense in which ethics are discussed critically by David Hoy: "that actions are at once obligatory and at the same time *unenforceable* is what puts them in the category of the ethical." This notion of ethics excludes actions that are enforced, that are not "freely undertaken" (Hoy 2004, 184). In mainstream discussions of ethics, this distinction usually refers to actions that have not yet become a "policy" (or deontology) requiring "compliance" (an Ethics) and therefore require ethical reflection and choice from an individual. But Hoy also means that some issues essentially remain absent from the perspective of institutionalized Ethics, and paradoxically, that is what makes them "ethical" because they might require from the individual and the group an engagement: a sort of ethical resistance. For Hoy, such an engagement refers, for instance, to actions that support "ethical resistance of the powerless others."

I consider the ethicality of care in such a speculatively ethical way. Commitments to taking care are always performed—or benefited from—through interdependent attachments even if we are not forced to do them by a moral order or policy, even if we don't want them to be ethically labeled. As I have affirmed earlier, for humans—and many other beings—to be alive, or endure, something, somebody, must be taking care, somewhere. One might reject care in a situation—but not absolutely without disappearing. The obligations of care are, however, asymmetrical. Speculatively thinking: when we commit to care, we are in *obligation* toward something—such as worms—that might have no power to enforce this obligation upon us. In turn, worms and other beings do take care of our waste even if they don't commit intentionally to it. That relations are not reciprocally symmetrical doesn't make them less vibrant with ethicality. What makes someone feel ethically obliged to worms can only be found in the grounded transformation of everyday practices that ravel asymmetrical modes of mutual obligation. And yet in such doings circulate the possibilities of radical (i.e., rooted and grounded) more than human webs of as well as possible ethicality. If adding a moralizing layer to these doings won't do, it is because it is not normativity that makes caring obligation possible but rather the ongoing reentering into co-transformation that further obliges the interdependent web.

It is possible to say that care webs have no subjective origins and endings to settle in. The circulation of care preexist individuals. The notions of constraints, obligations, and requirements, though referring to contingent necessities, cannot be closed, not because of abyssal existential uncertainties but maybe because being drawn to caring commitments has something of an "immanent obligation" that gets reinforced as we engage. As Elisabeth Povinelli beautifully puts it, immanent obligations grow in an ongoing process of which it is difficult to state where the initiation came to:

a form of relationality that one finds oneself drawn to and finds oneself nurturing or caring for. This being "drawn to" is often initially a very fragile connection, a sense of an immanent connectivity. Choices are then made to enrich and intensify these connections. But even these choices need to be understood as retrospective and the subject choosing as herself continually deferred by the choice. I might be able to describe why I am drawn to a particular space and I may try to nurture this obligation or to brake away from it, but still I have very little that can be described as "choice" or determination in the original orientation. (Povinelli 2011, 28)

This is the immanent, and ambivalent, force of care, for the better and the worse—it is also probably why care makes us so susceptible to pervasive hegemonic moralities.

Pausing: It Was the Joy (to the Grasshawgs)

Looking for ways to discuss what I find singular in permaculture care ethics has taken me far from what drew me to this journey. I pause to ponder this. Though it's been a while, I remember some moments quite vividly, and it is mostly joyful memories that I have continued to foster.[11]

I'm in my mid-thirties, and I don't recall having my hands happily in the soil since childhood. I actually do not really recollect *enjoying* touching the moist dark soil without guilt, but no one around in this training tells us not to get dirty—doesn't one teacher right there have a t-shirt that says "Dirt First"? I've never heard people talk about the soil below our feet so fondly. I kept thinking of my favorite teacher in primary school, Mrs. Christy, who allegedly took off her shoes on a school trip after a downpour and—yuck!—walked into the mud. I didn't see this myself, so I don't know if it was true or schoolroom gossip. I wish I had. Though probably I would have been as shocked as the other city kids instructed not to get dirty. But today those reluctances are gone, and I'm having a jolly good time getting muddy.

A good time? Yet the background in the training is still that of ecologies on the verge of disaster. After all, this is Earth Activism, oppositional, revolted. It is also discussed by participants as healing time, supportive time, for worried people, tired people, angry people, precarious people—

environmentalists with no health insurance ("I want to become a farmer"—
"Believe me, you don't"). But yes, being in the fields and woods, doing this
work with the soil, the water, and plants, learning about more than human
working patterns and how to foster abundance, envisioning that we might
actually change something, one garden at a time. It all feels really good.
The affective feel that remained with me for a long time after this training
was a sense of renewal of collective hope and joy in the face of a frighten-
ing and often depressing world. Feeling burnout, anxiety, and exhaustion
didn't need to be the only way to care. This mood—beyond feeling good
and thinking that was OK—became crucial to a transformation of my engagements. Three ecologies need to support and feed each other: psyche,
collectives, and Earth. I sense that it was the way in which doing and joy
were cultivated together that had this effect. A privilege, I'm acutely aware,
to dwell in maintenance work as a joyful activity. All things *not* being
equal: I don't mean that care work is fun per se. But these experiences did
change my relation to the toil side of everyday care. They too enlarged the
frame in a particular way.

I remember well the day we were learning how to work with water in
landscapes. We are all lying on the floor, eyes closed, a gentle voice accompanied by soft rhythmic drumming is leading us into a trance that follows
the water cycle. I now know that this practice is a usual feature of Earth
Activist Training and also of some pagan camps. I remember still very well
that moment—and not only because I discovered that I was susceptible to
trance work. But because there "I" am, a drop of rain falling to the ground
following a voice that tells me I'm a water molecule in this ground, blending and passing by creatures, traveling through tiny guts, and finally joining others in a waterbed and resting until some force pulls me through
again. Of course this is all imagination: How would I know what it feels
to be water? I'm not a shaman. Be what it may, the water cycle trance did
something. I remember imagining-seeing that worm and passing through
it, and feeling fond of it all, and it all stayed with me, to realize the many
nonhuman ones living down there, and I had never thought about the soil
really being so alive.

I recall another day when we are stirring and mixing stuff, powders and
bits with water with a wooden stick in an old metal can. We all get a chance

to stir the cauldron and giggle hard. But we are carefully making "compost teas" to feed the soils. Following recipes precisely, according to a conception of soils as foodwebs—the trophic webs of beings that inhabit the soils—drawn from the work of a soil scientist, Elaine Ingham, aka the Queen of Compost. I take detailed notes of the recipe, and though I had deliberately put my research self on vacation, I can't help but make a mental note to check Ingham out. I wonder if this is a scientific recipe. Is this science for the people? Science for the worms? It would take another few years before I eventually went back to think about this. Composting, however, became my favorite part. It seemed simple and unpretentious, a doable way to contribute to the life of soils for a city dweller and so important at the same time. As life turned out to be, I didn't become a skilled grower—rather the contrary. So while I had a garden, I became good at composting and chatting with my worms. I touched them carefully, trying to exert the lightest of pressures and not to tear them apart while turning the compost (a tricky mission as anybody who has tried will know).

These experiences finally composted into a new research journey that led me to care for what is happening to the soil and our relations with it (see chapter 5). Povinelli's notion of being "drawn to" feels so accurate: none of these bits of experience can account for how and when this happened, how I was drawn down to this world below. I didn't "choose" to become affected by soils, by who lives in them, and what people do with them. But I do sense that this happened through an embodied immersion in collective doings that enacted an ethics and by continuously cultivating this experienced obligation as a joy—cutting from other relations, cares, and doings that initially draw me to the hills over Bodega Bay.

Permaculture Obligations and the Ethicality of Care

Permaculture care ethics can be read as ethical obligations that are recreated through everyday doings. Reading these ethics through feminist approaches reveals them as immanent, contingent, and situated articulations of as well as possible relations of ecological care. This brings me back to issues that opened this chapter: engaging with the ethicality of everyday doings within a politics of care and striving toward human decentered ecological relations.

The personal-collective. I have stressed the significance of "personal" everyday ethics, within a politics of the ordinary related to a collective rather than based on individual choices. No field of ethical thinking has focused more on the politics (biopolitics or not) of everydayness, of ordinary and mundane ways of living, than feminist ethics of care (Jaggar 2001). As discussed in previous chapters, my perception of care ethics as a doing is particularly influenced by feminist sociologies of care labors (Malos 1980; Precarias a la Deriva 2006, 2004) as well as political theories of care (Tronto 1993). Reclaiming the significance of historically neglected values developing in the misnamed "reproductive" sphere of living, feminists insisted on how everyday practices of caring in "private" realms are political. This move went against the traditional reduction of politics to public life addressed earlier. From this perspective, care is an ethico-political issue, not only because it is made "public" but because it pertains to the collective and it calls upon commitment. Personal lives are both affected by what a world values and considers relevant *and* transformable through collective action. Thinking of practices of everyday care as a necessary activity to the maintenance of every world makes them a collective affair. As such, when someone is in obligation to care for a child or an elderly person, or an animal, s/he is doing a job for a collective, not only her/his "self" perpetuation nor that of "one" family. In permaculture movements, where care for the earth is an inseparable doing from care of the personal, ecological interdependency is not a moral principle but a lived material constraint—required and obliged. Conceived as such, the obligation of care corresponds to a perception of its endurance and necessity in the contingent naturecultural relational webs of life and death composed of multilateral interdependencies, eschewing an understanding of care as a moral universal, imposed from an outside, a utilitarian rationalist contract or altruistic ideal.

Care as a doing. Care is a necessary practice, a life-sustaining activity, an everyday constraint. Its actualizations are not limited to what we traditionally consider care relations: care of children, of the elderly, or other "dependents," care activities in domestic, health care, and affective work—well mapped in ethnographies of labor—or even in love relations. Reclamation of care is not the "veneration of 'feminine values'" (Cuomo 1997,

126), but rather the affirmation of the centrality of a series of vital activities to the everyday "sustainability of life" that has been historically associated with women's lives (Carrasco 2001). This is an important aspect for thinking a naturecultural meaning of care ethics. We need an even more radically displaced nonhumanist rephrasing of Joan Tronto and Berenice Fischer's generic notion of caring than I already proposed above by expanding "our" world. We need to disrupt the subjective-collective behind the "we": care is everything that *is* done (rather than everything that "we" do) to maintain, continue, and repair "the world" so that *all* (rather than "we") can live in it as well as possible. That world includes ... *all* that we seek to interweave in a complex, life-sustaining web (modified from Tronto 1993, 103). What the "all" includes in situation remains contingent to specific ecologies and human–nonhuman entanglements. What counts is the "interweaving" of living things that holds together worlds as we know them, that allows their perpetuation and renewal—and even that which helps to their decay as we have seen with the example of worms' labor of composting. Acknowledging the necessity of care in more than human relations, not as all that there is in a relation, not as a universal connection, but as something that traverses, that is passed on through entities and agencies, intensifies awareness of how beings depend on each other. Moreover, as we saw in chapter 2, if care is a form of relationship, it is also one that creates relationality—as much as it cuts it, delineating (nonrelational) spaces where someone's care is not required or refused. Permaculture care ethics consider that humans are not the only ones caring *for* the earth and its beings—we are *in* relations of mutual care. But it is important to think that we are not connected in an abstract transcendent sphere but, as Thom Van Dooren puts it with Deborah Bird Rose, "everything is connected to something that is connected to something else" (and not to *all* something else) (Van Dooren 2015, 60). An a-subjective notion of care in more than human worlds that takes this into account requires the speculative ethical imagination to consider the many ways in which nonhuman agencies are taking care of many human *and nonhuman* needs, in specific relations of ethos-creation, as much as humans, not the Human, not Anthropos per se, but humans in worlds, who develop ways to contribute to an *as well as possible*, a wellness that, in turn, only takes meaning within ecological constraints vital to maintenance, repair, and the possibilities of thriving.

Remediating "neglect." Ethical obligation to care stands against "neglect." By neglect I mean what happens when the doings of care are not attended—not when care is not required or when things are better cared for by being left to be. Also, labors of care are often neglected because they are considered less important—domestic, as petty, sentimental or personal-oriented tasks—than the ones that emphasize the autonomy and independency of individuals; they are as undervalued as are those who accomplish them. This has an ethico-political translation. When caring is neglected, obligations of care call upon commitment to share troubles and burdens of the neglected. Again, this is not a moral abstract principle of solidarity but a doing that takes meaning and value within relational arrangements—practices, ecologies. In Part I, I approached how this commitment can be considered intrinsic to knowledge and technologies. In naturecultural permaculture practices, ethical obligation is embedded in practices for remediating the neglect of Earth's needs—including humans. As such, these ethics attract attention to the invisible but *indispensable* labors and experiences of Earth's beings and resources. The ethicality here is about making us care for what humans—most of us—have learned to collectively neglect. This responds to the dilution of ethical obligation: not everything is ethical, nor is the burden of care universal and homogeneous—all humans would be *per essence* the pastoral carers of nonhumans. The ethical obligation to remediate neglect is asymmetrical and historically situated: *today* it might involve more humans assuming the everyday responsibility to intervene in unbalanced worlds, to respond to a biopolitical situation in which ones are in measure to care for others who are in need of being cared for, and to acknowledge the care value of more than human agencies.

Care as affective concern. It is easier to see how care is a material constraint and an ethical obligation when we associate it with the necessary material doings that get us through the day. But what about care as an affective force, contained in the phrase "I care"—associated with love, the recognition that something is important, as well as responsibility and somehow "concern" for another's well-being. The material and affective are entangled in an ethical perception of care as something we do and feel.

But thinking care as a doing also changes the way we envision care as affective concern. Feminists have shown how much affective labor can be energy consuming and how it is even a commodity—for example, in customer care and other services (Hochschild 1983; Vora 2009a; Dowling 2012). In a world in which inequalities make of care a burden mostly carried by ones at the expenses of others, "to care" can be devouring for women and other marginalized carers. So feeling an obligation "to care" is more than an affective and moral state. It has material consequences for those who assume it—coercively or not. As I said previously, in permaculture practices the condition of sustainable collective caring is the maintenance of resources, including those of one's energy. This is also why cultivating joy is part of the doing. In a conception of care as a collective good, care has to be shared, distributed, the "surplus" of life and energy that it produces returned to the carers in order to avoid affective and material burnout—including burnout of nonhumans subjugated in relations of ecological "service" and humans bound to the logics of productivist exploitation of nature (such as agricultural workers).

Care as situated. Ecofeminist philosopher Chris Cuomo has pointed out problematic assumptions in simplistic reclamations of care ethics for the natural world, in particular its reduction to purportedly "feminine values," interest in the concrete rather than the "abstract," nurturance, intimacy, ego denial (Cuomo 1997, 127). Cuomo pointed out two problems in these assumptions. First, from a feminist perspective, we cannot forget that automatically associating women as those in charge of these qualities is part of the oppressive systems that neglect caring. More generically, acknowledging this means facing that there are situations in which we could willingly abstain from giving care not only because it is good for others to be let be—like in cases of leaving an ecological relation to follow a better-without-humans course—but in order to refuse that care remains assigned to the same collectives. Second, and correlatively, she calls us to keep at heart that "the meanings and ethical relevance of acts of caring and compassion are determined by their contexts and objects" (130). Care is a necessary activity, but its actualizations are always relationally specific. Affirming this necessity does not imply universality. In every context, care responds to a situated relationship. On the ground, doings are always more "messy" than they appear in principles. As I said before, in permaculture, "it depends": it's a nuance that accompanies the *ways* acts of care are realized, and this again is affected by relational constraints—requirements of an ecology, obligations of the practitioners, *and* their struggles.

Noninnocent care. Far from being an innocent activity, care in nature-cultures cannot be purged from its predicaments: for example, the tendency to pastoral paternalism, the power it gives to care takers, and the unequal depletion of resources it implies in existing divisions of labor and exploitation of nonhumans and humans. In some contexts, care is inseparable from killing: like in weeding one's garden to make possible more fertile growth. As Haraway puts it, interspecies living is also about "mortal relatedness" (Haraway 2007b). Engaging with the ethicality in these fraught questions, Haraway argued for refusing to make anything and anybody "killable." Sometimes the question of how to care might mean that we have to engage with issues concerning if, why, and how to kill and for what: for example, in preoccupations with the welfare of the animals slaughtered for feeding (Evans and Miele 2012). And there will not be an easy answer. All these reasons and more confirm that care is not about ideal "good feel" relationships, something particularly crucial to think within the context of contemporary ecological engagements in shattered and disproportionately distressed geographies of naturecultures. Obligations of caring in naturecultures cannot be reduced to "stewardship" or "pastoral" care in which humans are *in charge* of natural worlds. Such conceptions continue to separate a human "moral" subject from a naturalized "object" of caring. Nor need we go to the other extreme: diluting the thinking of specific obligations of care in situational relations with nonhumans (or worst, in a *naturalized* conception of the *bios* collective—we are all just animals, period). These are poor generalizations that avoid engaging with actual situated naturecultures and the speculative efforts demanded from ecological thought and practice.

Alterbiopolitics

Eco-political awareness of the distressed state of Earth ecologies and its "resources," in a context of extension of consciousness about naturecultural catastrophe and massive extinctions, gives an acute meaning to the permaculture principle of returning the surplus rather than continuing in perpetual relations of extraction. It also gives further significance to re-thinking a naturecultural politics of care in times that are deeply antieco-logical, and in many ways anticollective. That good care is not granted by moral intention can be reaffirmed in this context by going back to the haptic trope, and in an earthy, permaculturish way: hands in dirt. It also brings back a fundamental aspect of the personal-collective permaculture ethics, that they are inseparable from a specific version of care as a poli-tics. Ethical doings in permaculture care ethics make a difference at the heart of biopolitics because they are alterbiopolitical interventions. "Alter" refers here to the embeddedness of the permaculture movement in alter-globalization strategies and struggles (Starhawk 2002)—that is, movements that affirm and engage with insurgent world-webs of life and possibility against colonial, ecocidal, capitalocentric predominant logics in the logics of globalizing and transnational network extension. "Alter" refers to a way of confronting biopowers by creating different forces of world-making relationalities—that would, in the words of Starhawk, cultivate "power-with" and "power-from-within" rather than "power-over" (Starhawk 1987; 2002). It would be fair to note that in bringing the ethics of permaculture to intervene in discussions about ethical involvements in biopolitics I have displaced the context of biopolitical intervention and ethical debate—as these are not the usual politics that mainstream biopolitics has been put in conversation with (with some notable exceptions that address biopoli-tics in a radical naturecultural meaning closer to the one I attempted here [Shiva, Moser, and Third World 1995; Esposito 2008]).

And yet it is these displaced connections that have led me to see how bio-political theorizations of new forms of ethics expose a recalcitrant focus on a humanist and individualistic body, however critical and politicized: the body-self, the body-citizen, or that of a "public" concerned about our bodies. Under contemporary conditions of pervasive forms of biopower and in the current worrisome state of planetary *bios* we are all dealing with fears, risks, rights, and protections in order to pursue the self-preservation of our own biological life. Individuals may or may not join in collectives but the prevalent understanding of ethics in biopolitics starts from how indi-viduals transform their lives and practices in resistance or in adaptation to the violences of biopower as a given. As such they are a compatible version of the Foucauldian *souci du soi*, a care of the self that though it is not separated from its collective inscription does start from cultivating a healthy relation to the self in order to live ethically. Ethical agency here is focused on practices that have as its purpose the edification of an ethical self. But, as I have argued, in Ethics hegemonic, this conception of ethical agency is not easily distinguishable from the contemporary imperative of self-care and its anticollective stance.

The alternative forms of biopolitical care I have addressed in this chapter do not start from or aim at "our selves," but neither do they put others before our selves. Care is embedded in the practices that maintain webs of relationality and is always happening in between. This meaning spreads the meaning of the ethical to the whole of a situation—to the agencies, materialities, and practicalities involved in the processes of caring. Here, the focus is not so much on the subjects of the so-called ethical action and decision-making but on how an ethos is fostered through relations and doings. Thinking this has several consequences. The affective potency of care is radically embedded on relationality and thus, crucially for my purposes, it is not to be *controlled by* a "subject" or one power source. Ethics is not a matter of applying moral principles by a subject to a senseless, and soulless, "material" universe: ethicality in the making resides in messy, muddled, concrete situations in which *an obligation of care* becomes at stake. Likewise, this makes it rather odd to think care as shaped by moral *control over* uncaring subjectivities. Caring happens more as a plane of "continuous experience," involving a range of lived material elements in decentered and multilateral relationships, rather than as a product of a bounded subject (Stephenson and Papadopoulos 2006). In the specific communities collected around permaculture ethics, we perceive care as it is continuously reenacted in inseparable entanglements between what is "personal"—how one individual is affectively engaged in attachments—and what is "collective"—a web of compelling relations, with humans and nonhumans, included in a community of practice in situations.

Relations foster care for some things rather or more than for others. In other words, acts of caring are never isolated, we care in an entangled way with what a specific situation requires care from and lures care for but this doesn't mean that what we care for is predetermined by given conditions. If to care is to be attracted, to be entangled with the recipients of (our) care in a relationship that not only extends (us) but obliges (us) to care, then *a world is being made* in that encounter that rather than determining (us), shifts (our) priorities. There is nothing *before* care that comes to be determined by it: rather, as we saw in the earlier discussion of Haraway's thinking with care, if "reality is an active verb," more than human realities have care hardwired in it. Within this extended conception of the ethical as ethicality and the assumption of the centrality of care in the very possibility of relating, the cares that oblige us can neither be uniquely explained by the contexts of forces and interests that constrain nor be abstracted from them. When we think about what we care for: one moment it seems it would be easy to remove our care; the moment after we realize that our care does not belong to us, and that that/whom we care for, somehow *owns* us, we *belong* to it through the care that has attached us.[12]

My hope is that this conception of care opens ways to think a decentered circulation of ethics in more than human worlds. Care as a doing and as ethos that creates ethical obligation does not need to be primarily directed to the ethical edification of human selves: it is about doings required by living communities to live as well as possible. Living in nature-cultures requires a perspective on the personal-collective that, without neglecting human individual bodies, doesn't start *from* these bodies but from awareness of their more than human interdependency. This requires a decentered perception of *bios* involved in sustaining these relations, an ethics that includes nonhumans responsibly but in nonexceptionalist, nonpaternalistic ways as belonging to this living community. This way a bodily ethics in biopolitics is not only about more awareness of how politics increasingly shapes the biological, corporeal, dimension of "our" existence, but about how to better cultivate our belonging to bios as a form of living community that goes beyond "our" existence (Esposito 2008). In a naturecultural world in which politics and ethics conflate in biopolitics, alterbiopolitical interventions are about working within bios with an ethics of collective empowerment that puts caring at the heart of the search of transformative alternatives that nurture hopeful thriving for all beings.

With its ethical explorations, this chapter stirred this book's journey toward an ongoing collective reimagination of ecological existences that focuses less on coping with biopower, adapting or resisting, and more on creating alternative forms of collective and caring politics within bios. The next and final chapter of this book probes further into the speculative possibility of altering human-centered conceptions of the webs of ecological care, working with a notion of earthy living collectives that encompass human and other than human agencies beyond idealized, bifurcated ideas of nature and exceptional humanity. These are not easy things to think or to do, but they are vital. Chapter 5 focuses on how human–soil relations are being transformed in an atmosphere of urgency about the neglected state of planetary soils. In a tense relational field where the future sways between hope and doom, I focus on scientific approaches to soil as living and on ecological practices of soil-care that could be altering the dominant conception of soil as a resource for human consumption, opening to conceiving soils as communities of kin.

❖❖❖❖❖❖❖❖❖❖❖❖❖❖❖❖❖❖❖❖❖❖❖❖❖❖❖❖❖❖❖❖❖❖❖

five

Soil Times

The Pace of Ecological Care

Human-soil relations are a captivating terrain to engage with the intricate entanglements of material necessities, affective intensities, and ethico-political troubles of caring obligations in the more than human worlds marked by technoscience. Increasingly since the first agricultural revolutions, the predominant drive underlying human–soil relations has been to pace their fertility with demands for food production and other needs, such as fiber or construction grounds. But at the turn of the twenty-first century, Earth soils regained consideration in public perception and culture due to global antiecological disturbances. Soils are now up on the list of environmental matters calling for global care. The Food and Agriculture Organization of the United Nations declared 2015 the "International Year of Soils," expressing concerns for this "finite non-renewable resource on a human time scale under pressure of processes such as degradation, poor management and loss to urbanization" (FAO 2013). Soils have become a regular media topic, drawing attention to the "hidden world beneath our feet" (Robbins 2013), a new frontier for knowledge and fascination about the life teaming in this dark alterity. Human persistent mistreatment and neglect of soils is emphasized in calls that connect the economic, political, and ethical value of soils to matters of human survival. Recent headlines by environmental analysts in the UK press reiterate this: "We're Treating Soil Like Dirt. It's a Fatal Mistake, as Our Lives Depend on It" (Monbiot 2015) or "Peak Soil: Industrial Civilisation Is on the Verge of Eating Itself" (Ahmed 2013). Warnings proliferate against a relatively immediate gloomy future that could see the global exhaustion of fertile land with correlated food crises. So while soils remain a resource of value extraction for human consumption and a recalcitrant frontier of inquiry for science, they are also increasingly considered endangered living worlds in need of urgent ecological care.

I end this speculative exploration of meanings of care in more than human worlds by beginning another research journey, grounded in the specific landscape of ecological care for the soil. In this book I have worked from and for a vision that embeds care relations in mundane doings of maintenance and repair that sustain everyday life rather than on moral dispositions. It is partly because of the devalued significance of care that feminist research on practices of care is often oriented by an ethico-political commitment to investigate the significance of neglected things, practices, and experiences made invisible or marginalized by dominant, "successful" (technoscientific) mobilizations. This way, paying attention to practices of care can be a way of getting involved with glimpses of alternative livable relationalities, with other possible worlds in the making, "alterontologies" at the heart of dominant configurations (Papadopoulos 2011). In this spirit, the critical inquiry into human–soil relations of care presented in this chapter is not so much driven to debunk the productionist subjection of soils but by an aspiration to engage speculatively with imperceptible tendencies that could be troubling and reworking these dominant relations from within by transforming everyday soil care. So, like all the other chapters, this one is written from the partiality of a speculative commitment: thinking with care as a way to elicit conceptions and practices that have the potential to disrupt the reduction of soil to a resource for humans. Attention to ways in which notions of "soil care" could potentially be transformed in these times of environmental unsettledness brings to light possible alternative practical, ethical, and affective ecologies. I therefore engage with soils as matters of care: human–soil relations of care and soil ontologies are entangled. What soil is thought to be affects the ways in which it is cared for, and vice versa, modes of care have effects in what soils become.

In this chapter, the speculative thought on more than human care follows my research immersion into contemporary changes in human–soil relations, to which I was drawn through my interest in permaculture. But by engaging care to think a context of fearful futures around the destruction of Earth's soils and by looking into current challenges to the longstanding confirmation of productionist relations, an additional theme I was not expecting also came forth: the uneasiness posed by care within anthropocentric temporalities of technoscientific futurity. Care, of course, has been traditionally associated with "reproducing" rather than "productivity," and so thinking speculatively with "care time" where productionist timelines prevail still opens interesting questions. This book therefore ends by inviting attention to a temporal dimension of care that the previous chapters only implicitly hinted at—care as the fostering of the endurance of objects through time (maintenance against breakdown), haptic care for the imperceptible politics of the everyday (rather than the irruption of events). By exploring this elusive but important feature of the doings of care—that is, the recalcitrance of the temporality of care to productionist rhythms—I reflect on how care time entails "making time" to get involved with a diversity of timelines (such as the ones involved in the living soil) that make the web of more than human agencies. Human–soil relations have a complex and fascinating history with very different local actualizations. My focus here is on their contemporary history in traditions driven by industrial agricultural revolutions—rather than exploring alternative relations with soil in other cultures and geographies. This is partially because this is where I am situated, and where my research is still at the time of writing, but also because I am looking for indications that the traditional harnessing and subordination of multiple temporal rhythms of soil care to the linear temporalities of technoscientific productionism could be contested within the very inheritances of agricultural revolutions. So, as in previous chapters, I'm not looking to create a space for care outside present predicaments and hegemonies, but within.

My point of departure is soil science, a scientific field that has been closely intertwined with societal and economic concerns across the decades, a relation deeply affecting its research agendas. While the importance of soils for agricultural needs has bound knowledge about soil to human economies of survival since ancient times, it is only since the mid-nineteenth century that scientific developments in chemistry, physics, and biology coalesce into the interdisciplinary field of the soil sciences. It is in close entanglement with this modern history that the soil became a distinguishable object of scientific, experimental, and field studies, applied and nonapplied. This is a fascinating field that is changing fast. And part of these changes are disturbing the traditional relation with productionism and exposing soil as a living world rather than a mere receptacle and input for crop nutrition.

Yet in order to indicate how something might be changing, I first delve into the historical context that gives significance to discussing current scientific conceptions of soil's ecological interdependencies in terms of their temporal implications. I then specifically focus on one of those conceptions, the "foodweb" model of soil ecology, which conceives soil as a living multispecies world. In order to think speculatively on this vision's potential to alter conceptions of soil care beyond science, I explore how it has become a figuration of embodied caring human–soil relationalities across different spheres of soil practice. The significance of the foodweb conception goes beyond its explanatory power, or epistemic value for science, to engage humans into eco-ethical obligations of care. From a temporal perspective, these obligations require an intensification of involvement in *making time* for soil-specific temporalities. Focusing on the temporal experiences of ecological care helps to reveal a diversity of interdependent temporalities of beings and things, human and not, at the heart of the predominant futuristic timescales of technoscientific expectations. It is technoscientific futurity that care time might intercept, because getting involved with soil's temporalities in a more caring way implies a disruption of current modes of temporal dominance in more than human worlds, including their ratification by prevalent conceptions of innovation.

Technoscientific Futurity

Human agricultural practices have exhausted soils across the world well before industrialization (Hillel 1992), pushing human populations to leave depleted soils behind in search of fertile grounds. In the current global productionist regime, options are recognized to be narrowing, as the extension of agricultural land by forest clearing is a documented factor of climate change, and the intensification of production in available land is destroying the resource. Humanity's vital need for soil serves the argument that the acceleration of its loss might be more worrying than the well-reported peak oil (Shiva 2008; Wild 2010). "Peak soil"—and the correlatives "peak nitrate" and "peak phosphorus"—refer to forewarnings of "economic" collapse by which a resource is bound to exhaustion without equivalent efforts to renew as it "becomes more difficult to extract and more expensive" (Dery and Anderson 2007). Countless accounts refer to strains on soil caused by human population growth, warning of famine waves, typically reciting figures approaching ten billion by 2050, announcing famine outbreaks if action is not urgently taken to ensure food security. And yet soil exhaustion is also blamed across the board on industrialized and unsustainable forms of agriculture, and so many see intensifying food production through technoscientific innovations as a misled perilous response to food security (Tomlinson 2011; McDonagh 2014). Similar to other environmental warnings, such as urging people to "Wake Up, Freak Out—Then Get a Grip," in response to climate change "tipping points,"[1] the temporal emergency in soil breakdown warnings is clear: the time to care more and better for soils is *now*.

It is likely that the impending loss of soil will affect how the inheritors of agricultural revolutions care about this vital universe. And what this could mean is also marked by tensions in this temporal atmosphere. The future of soils appears to be pulled forward by an accelerated timeline toward a gloomy environmental future, while the time left for action in the present is compressed by urgency. And so the temporal pace required by soil's ecological care as a slow renewable resource might again be at odds with these conditions of emergency, running against the accelerated linear rhythm of intervention characteristic of technoscientific futuristic response, traditionally straddled to a productionist pace. Sensing technoscientific futurity as a specific "timescape," a notion borrowed from Barbara Adams's sociology of time, "stresses the temporal features of living. Thinking with timescapes, contextual temporal practices become tangible. Timescapes are thus the embodiment of practiced approaches to time" (Adam 1998, 10). Timescapes are devices to think epochal time, in terms of their everyday actualizations, resistances, and contradictions. In other words, epochal, practical, and embodied timescales are entangled; they do and undo each other. Thinking technoscientific futurity as a timescape allows space for thinking practices as time-making and envisioning what doings and agencies could be disrupting the overpowering atmosphere of ecological anxieties so consistent with the hegemony of future-oriented timelines in technoscientific societies.

The features of this particular timescape of futurity have been illuminated in science and technology studies and sociology from several critical perspectives. First, technoscientific futurity has been discussed with regard to the persistence of a modern paradigm that associates the future with progress, with an ethico-political imperative to "advance" that remains solidly the orientation of linear, "progressivist," timelines—while the past acts as a discriminatory signifier of development delay (Schrader 2012; Savransky 2012). From the perspective of this hegemonic timescape, as faith in modern linear progress is increasingly put into question by an environmental crisis, uncertainty prevails—and catastrophic regression seems inescapable (Beuret 2015).[2] Second, the future orients practices. It acts as the inexhaustible pull of technoscientific "expectation"—that is, the socio-affective engine of innovation-driven political economies (Brown and Michael 2003; Hedgecoe and Martin 2003; Borup et al. 2006; Wilkie and Michael 2009)—as well as of "promissory" science (Thompson 2005). Here technoscientific innovation is situated and affected by a shared timescape of futurity typical of late capitalist economies, a timescape that fuels "preemptive strategies" and subjects practices in the present to a productionist ethos increasingly committed to the speculative extraction of future economic value (Cooper 2008; Papadopoulos, Stephenson, and Tsianos 2008b; Dumit 2012; Papadoupoulos, Lilley and Papadopoulos 2014). Third is the "anticipatory" affective state of technoscientific futurity that Vincanne Adams, Michele Murphy, and Adele Clarke have insightfully characterized as one of permanent anxiety, "in which our 'presents' are necessarily understood as contingent upon an ever-changing astral future that may or not may be known for certain, still must be acted on nonetheless" (Adams, Murphy, and Clarke 2009, 247). Technoscience's innovation-driven focus on novelty fosters uncertainty and expectation about an imminent breakthrough that could change it all for the better or the worse. Any meaningful act in the world of promissory capitalism involves taking risks and acting fast.

In this form of futurity the everyday experience of time is one of permanent precariousness: an ongoing sense of urgency and crisis calls us to act "now," while the present of action is diminished, mortgaged to an always unsure tomorrow. Industriously advancing and producing might give the beat to get practice going, but the continuity of existence is also constantly challenged, injecting drama and fear into everyday doings. The "hype" (Brown 2003) characteristic of futuristic progress-driven innovation is codependent with fear of doom and hope for salvation (Haraway 1997b; Kortright 2015). The restless work involved in managing anticipation and calculation (Clarke 2016) in the face of uncertain futures is the late capitalism pendant of modernity's impossible efforts to manage and control time (Adam 1998).

The three lines of critique outlined above characterize different scales, albeit intimately entangled, of a dominant mode of futurity in technoscience: the temporal frame of an epoch still marked by a linear imperative of progress versus fears of regression; the time embedded in practices paced to a productionist ethos; and the experienced, embodied time of restless futurity. What these analyses of temporality show is that the future is crucial in "constituting" the present of everyday life in technoscience (Michael 2001). They also expose, and somehow ratify, the intrinsically futuristic character of dominant notions of technological and scientific innovation. Yet there are also motivations to question our ambivalent enthrallment with the future.

Thinking from the specificity of timescapes intercepts temporal determinations. Sociohistorical critiques of temporality show how different societies and epochs foster and promote different experiences of time. Conversely, looking at temporality from the perspective of everyday experience, time is not an abstract category, or just an atmosphere, but a lived, embodied, historically and socially situated experience. Time is not a given; it is not that we have or not time but that we *make it* through practices (Dubinskas 1988; Whipp, Adam, and Sabelis 2002; Frank Peters 2006; see also Wyatt 2007). Temporality is not just imposed by an epoch or a dominant paradigm but rather made through sociotechnical arrangements and everyday practices. If we want to think the possibility of a diversity of practices and ontologies, the progressive, productionist, anticipatory temporal regime, although dominant, cannot be the only one, nor is it exempt from coexisting with other timescapes, as well as it comports tensions within a variety of timescales that comake and might contest each other.

The case for exploring (and enacting) alternative temporalities is made even more compelling by a renewed emphasis on temporal diversity in the social sciences and the humanities. Especially relevant for this chapter is interdisciplinary work marked by an ecological critique of linear and anthropocentric temporalities (Bastian 2009). Indeed, a diversity of eco-temporalities is revealed when multispecies, more than human, scales are considered (Schrader 2010; Choy 2011; Bird Rose 2012). These insights are of specific importance to research on human–soil relations and ontologies. Soil is created through a combination of the long, slow time of geological processes such as those taking thousands of years to break down rock— that Stephen Jay Gould qualified as "deep time" (1987)—and by relatively shorter ecological cycles by which organisms and plants, as well as humans growing food, decompose materials that contribute to renew the topsoil. Both micro and macro timescales at stake in ecological relations involve different time-frames than those of human lifespan and history (Hird 2009). This is not only a philosophical or scientific problem, it is an ethi-cal and political one. In the words of Jake Metcalf and Thom Van Dooren, attending to time as materially produced, as lived time, draws attention to "ruptures in ecological time." This requires thinking of timescapes that could be "liveable for humans and non-humans alike" (Metcalf and Van Dooren 2012, v). This is a crucial task today, they affirm, when "eco-logical well-being depends on aligning the temporal dimensions of many beings, and the consequences of disruption and slippage between times" (vi). The emphasis on temporal diversity has implications for how we live together and how we belong in communities, that is, in creating "temporal belongings" for humans and nonhumans (Bastian 2014).[3] Whether we name this Epoch the Anthropocene (Zalasiewicz et al. 2011) to emphasize the impact of human technoscientific progress, or the Capitalocene (Moore 2014) to reflect the capitalist politics of *some* humans, drawing attention to the entanglements and frictions within more than human experiences and timescales has ethico-political, practical, and affective implications (Haraway 2015).

Finally, engaging with different ways of experiencing time could have additional significance for the way that we look at the temporality of science and technology. In particular, making time for care time might disrupt the "imaginings of technology" that, as Steve Jackson (2014, 227) has suggested, keep the language of innovation for the new "bright and shiny" and for quasi-teleological achievements "at the top of some change or process." I will be discussing how contemporary approaches to soil care disrupt this vision of innovation. Here attending to care and questioning innovation join into interrogating the "productivist bias" —which Jackson also identifies at work in science and technology studies and calls to question (see also Papadopoulos 2011, 2014b). Here a feminist politics of care in technoscience—akin to Jackson's and others' attention to practices of "maintenance" and "repair" (Denis and Pontille 2014)—appears particularly relevant. It is therefore more generically that I'd like to explore how the temporality of care, of soil care in this case, offers an inquiry into different modes of "making time" by concentrating on experiences that are obscured or marginalized as "unproductive" in the dominant futuristic drive.

Focusing on experiences of soil care that offer alternative modes of involvement with the temporal rhythms of more than human worlds can contribute to disrupt the primacy of technoscientific futurity by acknowledging temporal diversity and questioning the anthropocentric traction of predominant timescales and notions of innovation. To begin, I situate the relevance of a discussion of soil-science knowledge with regard to matters of temporality by highlighting contemporary tensions around the future of soil science and how it can contribute to the mattering of soil in an epoch of ecological dislocation.

Soil Science Futures and Epochal Emergency

Soil science is a relatively young discipline that only emerged as a distinctive field in the mid-nineteenth century, when developments in chemistry, physics, and biology combined with research agendas driven by concerns around food production. Yet, until recently, the most important accounts of the discipline's history had been written by scientists adopting a classic "internalist" perspective, addressed to soil scientists, and focused on central scientific figures, paradigms, and conceptual shifts (Krupenikov 1993; Yaalon and Berkowicz 1997). Engaging properly with this complex history goes well beyond the purposes of this book. What is important to mention is how few and scattered are the indications in this literature of the entanglement of scientific developments with socioeconomic contexts, let alone of connections with agricultural capitalism (Moore 2010). Jean Boulaine notes how the first agricultural revolution in seventeenth-century Britain was fueled by the introduction of off-site natural fertilizers first extracted and imported from the colonized Americas. As these resources became exhausted, fertilizers were developed artificially, propelling soil chemistry through its contribution to industrial manufacturing (Boulaine 1994). Only recently have discussions about the future of soil science in the past twenty years been paired with an interest in historical accounts of the discipline and in understanding its relation with wider socioeconomic contexts (Bouma and Hartemink 2002). Looking at the entanglements between advancements in the field and moments of crisis affecting soil as resource could contribute to these efforts.

One famous example is the "dust bowl" phenomenon in the 1930s, by which powerful windstorms carried away the topsoil of intensively farmed land, devastating livelihoods and leading to the displacement of hundreds of thousands in the North American high plains. Environmental historian Daniel Worster (1979) showed how this disaster, which still marks the imagination of environmental devastation in the United States and beyond, brought with it an intensified wave of technically enhanced soil exploitation based on agrochemical inputs and innovative irrigation systems. Douglas Helms, historian of the U.S. Soil Conservation Service, shows how the dust bowl had an immediate effect on scientific and social investment in soils, including an increase in public support of U.S. soil conservation policies and the extension of soil surveying and mapping enterprises (Helms 1997).[4]

Another well-known example is how, in the late 1950s, anxieties about an ever-expanding population and imminent famine, particularly in Asia, contributed to public support for the technoscientific complex that set in motion the so-called Green Revolution, accomplished by combining artificial fertilizers, newly developed high-yield seed stocks, and chemical pesticides, leading to intensive cultivation and unprecedented yield. Today, controversies persist about the social and environmental effects of the Green Revolution (Cleaver 1972; Shiva 1991; Thompson 2008). The dramatic consequences for farmers of the destruction of soils and water that followed this wave of agricultural intensification still gather public attention (Weiss 2012). However, the attraction of a new Green Revolution to respond to current threats to future food security has not faded. It remains a model to "unlock the potential of agribusiness" in the untapped lands of the African continent (World Bank 2013); the concept is kept alive in scientific circles in reformed, more "sustainable" versions (Sánchez 2010, 2004), often turning attention, though without scientific consensus, to the power of genetically modified crops that could cope with impoverished soils.

Historically, social emergency and gloomy uncertainties about soil resources and practices are not new to soil scientists. Fertility, erosion, pollution, nutrient depletion, and carbon capture are just some in the series of concerns that modern soil science has been called on to remediate. These instances in the history of human–soil relations also can be read in terms of how they expose a combination of anxious restlessness about the future—in the face of disasters such as the dust bowl or fears of mass famine—with ambitious responses based on large-scale innovations that confirm the technoscientific productionist drive. A posteriori, we can see how the effort of value extraction from the soil rarely has been tempered by disasters. In the current context the atmosphere of urgency and anxiety about imminent resource exhaustion seems to give impetus to accelerated extension of the promissory futures market networks around vital natural resources—thanks to new opportunities of exploitation sometimes even opened by environmental degradation, for example, oil extraction in newly accessible arctic zones (Johnson 2010). In the case of soils, these economic moves can be seen in the rush to grab fertile land (Borras et al. 2011):[5] the less there is left, the more valuable an investment it becomes, and its intensified exploitation is further accelerated.

Contemporary concerns in soil science around the historic and sociopolitical role of the discipline can be read against this background because today science is called to mobilize in a context of global ecological change and possibly disaster to address pressing concerns around the state of soils and their capacity to provide.[6] This is not the only reason why soils are "back on the global agenda," but it does contribute to a "renaissance" of soil science as a privileged way of responding to the crisis of soils (Hartemink 2008; Hartemink and McBratney 2008). In contrast, the scientific identity of the field is put at stake. Soil physicist Benno Warkentin asks: "Can we ensure that soil science as a discipline is not lost in the coming competition of responses to society's needs?" So while the applied character of soil science seems uncontroversial, there are arguments to preserve a "basic" value to soil science: a focus on responding to societal demands could result in a potentially hazardous "technology fix" (Churchman 2010, 215). In a context where the social relevance of science has become difficult to disentangle from the industrial ventures of promissory capital, the calls to keep a fundamental role for science might be losing their traditionally conservative connotations—to keep science "pure"—to become closer to an understated mode of resistance.

Alfred Hartemink, a scientist who has dedicated considerable efforts to promote engagement with the discipline's history and future, nonetheless reminds the entanglement of the scientific enterprise with an imperative to look into the future when he states: "For any scientific discipline it is good to look back and make out what has been achieved, how it was done and whether anything can be learned from the past. No doubt that is a respectable activity but it will not yield scientific breakthroughs. If you want to stay in business as a science it is healthier to look forward" (Hartemink 2006, vii). Perhaps more than any other modern social practice, science is actively and performatively embedded in the progressive, promissory, productionist epochal timescape. In particular, modern science's inherent progressivism reacts against any notion suspected of "turning back the clocks." As described in the previous section, within such a conception progress is either valued for its gains or feared and blamed for its repercussions. Advances in science can be questioned but not a general ineluctable progression to the new or to a "breakthrough." In other words, in the epic narrative of scientific mobilization that Isabelle Stengers identified as core to modern science's social identity (Stengers 1993), either we go forward or backward. Yet in spite of the traction of epochal futurity for science, debates and tensions about soil science's future reveal frictions. One important theme around which these tensions can be shown to crystallize today is the challenge to increase agricultural yield while promoting sustainable soil care.

Hartemink's words above are extracted from *The Future of Soil Science*, a volume he edited in 2006 for the International Union of Soil Science (IUSS) in which a number of influential soil scientists around the world were asked to share their thoughts about trends and directions in the field. These interventions expose tensions in how soil scientists see the future of the field at this particular time. Reflecting on the future of their science, some hold to an inherently progressive vision:

> While doomsayers expressed apprehension and pointed fingers, soil scientists, along with plant breeders and agronomists, ushered in the green revolution by enhancing agronomic production by growing input responsive varieties on fertile and irrigated soils. As has been the case in the 20th century, those holding neo-Malthusian views will again be proven wrong through adoption of recommended management practices for sustainable management of soil resources. (Rattan Lal in Hartemink 2006, 76)

Just as soil science participated in the Green Revolution epic and enhanced production, it can participate in the grand enterprise the world is facing to feed itself with more sustainable practices that would avoid previous pitfalls by which "alas, impressive gains in food production in the twentieth century were achieved at the cost of environmental quality" (Rattan Lal in Hartemink 2006, 76). Science can continue going forward, as usual: strains on the environment by increased demand are not necessarily (framed as) conflictive but are part of a progressive history of accumulated wisdom. Of course, what the way forward could be with regard to the damage of past technologies is what remains at stake. As another scientist states in the same publication:

> Well-fed industrialised countries with stagnant or declining populations might have the luxury of such notions, however valid they might be for their conditions, but . . . developing countries cannot—and should not—be lulled into the mistaken belief that they can get by without the use of fertilizers. (John Ryan in Hartemink 2006, 123)

This cautionary posture reveals the embeddedness of scientific discussions in tensions and controversies around the prospect of another Green Revolution, as the way forward, if we care about the fate of the great majority.

The seamless vision of the environmental leadership of soil science is further problematized by emphasizing its integration with socioeconomic requirements. For Dick Arnold, "Soil science operates simultaneously in the realms of ecology and of economics, each of which marks time by different clocks," and so the future of soils depends on how economics/society will trade off between sustainability and exploitation (Arnold in Hartemink 2006, 7). Here a different underlying narrative implies that an ecological soil science will follow an ecologically progressive society, in which "opportunities are golden for imparting the knowledge and wisdom of soil science" (8). Yet more pessimistic are those who recognize a historical failure of soil scientists to convince agronomists of ways to produce without damaging the environment, something that French soil scientist and former president of the International Society of Soil Science Alain Ruellan emphasized in his critical review of this volume (Ruellan 2007).

The field of soil science is vast and transdisciplinary, and the variety of scientific voices shouldn't be reduced to the dynamics and tensions I am delineating here. Across the contemporary literature that addresses the societal role of soil science, most scientists associate the future of the discipline with a commitment to sustainability. And it is to this purpose that I am attempting to contribute by illuminating the trends displacing dominant logics in soil exploitation. So what can be learned by illuminating tensions around the future? I believe that it is important to examine the assumption of an alignment of soil science with an ecological temporality—as it was oriented by a clock somehow "naturally" marking a different time than unsustainable "economics" (or the "social"). This obscures how not only economics but also science has been resolutely oriented by a typically linear orientation to the future based on producing output and profit through innovation. If the productionist logic is not moderated but, rather, accelerated, in times of anxious futurity, and if, as I have argued, techno-scientific progressive temporality is intrinsically entangled with productionism, the alternative seems bleak or "infernal" (Pignarre and Stengers 2011): intensify agricultural gain (and further exhaust soils) or the world will starve. Environmentally concerned scientists will have to find ways to work within pressures unlikely to go away. And while at the level of epic scientific mobilization it remains difficult to disentangle science from technoscientific futurity, there are conceptual and practical reorientations within soil science that could be troubling this temporal alignment from within, by displacing the productionist ethos that subjects soil care and, more generally, human–soil relations to the extraction of future economic value.

From Productionism to Service—and Care?

Soil biologist Stephen Nortcliff speaks of a change in focus from research in the 1970s and 1980s, when sustainability concerns focused on "maintaining yield" rather than the "soil system": "How things have changed as we have moved into the 21st Century! Whilst maintaining agricultural production is still important the emphasis now is on the sustainable use of soils and limiting or removing the negative effects on other environmental components" (Nortcliff in Hartemink 2006, 105). Nortcliff is not alone. A disciplinary reassessment seems to be taking place. This could be a significant shift in the historical orientation of soil science, as summarized by soil scientist Peter McDonald:

> Soil science does not stand alone. Historically, the discipline has been integrated with all aspects of small farm management. The responsibility of maintaining good crop yield over a period of years was laid upon the soil. Research into soil fertility reflected this production-oriented emphasis during most of the nineteenth century . . . the focus of their efforts remained, and to a large extent still remains, to benefit overall harvests. (McDonald 1994, 43)

Guaranteeing yield through production is obviously an essential drive of the agricultural effort. But critical research on agriculture refers to *productionism* more specifically in terms of the intensification that drove agricultural reform in Europe from the seventeenth century onward. This culminated in the mid-twentieth century with the industrialization and commercialization of agriculture and the international expansion of this model through the Green Revolution's assemblage of machines, chemical inputs, and genetic improvements. In *The Spirit of the Soil*, philosopher of agricultural technology Paul B. Thompson argues for an ethics of production and summarizes productionism as the consecration of the aphorism "Make two blades of grass grow where one grew before" (1994, 61).[7] Critiques of productionism address the absorption of agricultural relations within the commercial logic of intensification and accumulation characteristic of capitalist economies. In other words, productionism is the process by which a logic of production overdetermines other activities of value (Papadopoulos, Stephenson, and Tsianos 2008b; Papadopoulos 2014b). Agricultural intensification is not only a quantitative orientation—yield increase—but also a way of life, and a qualitative mode of conceiving relations to the soil. While it seems obvious that growers' and farmers' practices, whether grand or small scale, pre- or postindustrial, would be yield-oriented, productionism colonizes all other relations: everyday life, relations with other species, and politics (e.g., farmers' subjection to the industry-agribusiness complex). The increasing influence of logics of productionist acceleration and intensification through the twentieth century can be read within scientific approaches to soil. One notable example can be found in chemistry's contribution to turning cultivation into a productionist effort. Soil physicist Benno Warkentin explains how early studies on plant nutrition were first based on a "bank balance" approach by which nutrients assimilated by plants were measured with the idea that these had to "be added back to the soil in *equal* amounts to *maintain* crop production." But the "balance" emphasis changed after 1940 with an increase in off-farm additions to the soil, bringing artificial fertilizing materials, external to a site's material cycles and seasonal temporalities, in order to bolster yield. The aim of this increase was to ensure "availability of nutrients for *maximum growth, and timing for availability* rather than on the total amounts removed by crops" (Warkentin 1994, 9, emphasis added)—that is, not so much to maintain but to intensify the nutrient input in soils beyond the rhythms by which crops absorb them. These developments confirm a consistent trend in modern management of soils to move from maintenance—for instance, by leaving parts of the land at times in a fallow state—to the maximization, and one could say preemptive buildup, of soil nutrient capacity beyond the renewal pace of soil ecosystems (Hillel 1992). This makes visible how the tension between production and sustainability at the heart of soil science involves misadjusted temporalities: between soil as a slowly renewable entity and the accelerated technological solutions required by intensified production.

   This is not to say that soil scientists—or even practitioners who live by the productionist credo—have not taken care of soils. Remediating worn-out soils has been at the heart of the development of soil science since its beginnings and was related to the socioeconomic concerns that influenced early soil studies (Warkentin 1994, 14). Numerous soil scientists have been committed to conserving soils and working with farmers to foster ways of caring for them while maintaining productivity: "soil care" is a notion widely employed (Yaalon 2000). Moves to interrogate productionism seem nonetheless to question conceptions of soil care in the light of a broader societal realization of the untenable pressures on soil. In science and beyond, the persistent productionist ethos overlaps today with an "environmental era" starting in the 1970s and influenced by a conception of environmental limits to growth that place "the living earth . . . in a central position" (Bouma and Hartemink 2002, 137). This has marked soil science—many researchers, for instance, pointing at the unsustainable destruction and deterioration of natural habitats associated with an excessive use of agrochemicals (134). Most sociohistorical accounts of the soil sciences since the early 1990s recognize this "ecological" turn: "in the present era of soil science . . . the questions are on a landscape basis, have an ecological nature, and ask about the sustainability of natural resources" (Warkentin 1994, 3–4).

   What can a critical analysis of the articulation of the temporality of productionism and relations of care contribute to these transformations? In a sense, there is an inherent ambivalence contained in these relations whereby the future is simultaneously hailed as central and "discounted," as Adam emphasizes with regard to short-term thinking that pushes to exploit natural resources today at the expense of future generations (Adam 1998, 74). And yet, the temporality of productionist-oriented practices in late capitalist societies remains strongly future-oriented: it focuses on "output," promissory investments (led by so-called agricultural futures), and on efficient management of the present in order to produce it. This is consistent with how, as described above, restless futurity renders precarious the experienced present: subordinated to, suspended by, or crushed under the investment in uncertain future outcomes. Worster's account of the living conditions of farmers who outlived the destruction of successive dust bowls to see the return of intensified agriculture and successful grand-scale farming are also stories of discontent, debt, and anxiety, echoing farmer experiences worldwide living under the pressures of production (Worster 1979; Shiva 2008; Münster 2015). So though the timescale of soil productionist exploitation discounts the future by focusing on the benefit of present generations, the present is also discounted, as everyday practices, relations, and embodied temporalities of practitioners embedded in this industrious speeded-up time are also compressed and precarious. Productionism not only reduces what counts as care—for instance, to a managerial "conduct" of tasks to follow (Latimer 2000)—but also inhibits the possibility of developing other relations of care that fall out of its constricted targets. It reduces care from a coconstructed interdependent relation into mere control of the *object* of care.

And it is not only human temporalities, but also more than human, that are subjected to the realization of this particularly linear timescale focused on intensified productivity. It could be argued that within the productionist model the drive of soil care has mostly been for the crops—that is, importantly, plants as commodifiable produce (which also begs the question of what kind of care is given to plants reduced to crop status). In the utilitarian-care vision, worn-out soils must be "put back to work" through soil engineering technologies: fed liters of artificial fertilizers with little consideration for wider ecological effects or made host for enhanced crops that will work around soil's impoverishment and exhaustion. In sum, soil care in a productionist frame is aimed at increasing soil's efficiency to produce at the expense of all other relations. From the perspective of a feminist politics of care in human–soil relations, this is a form of exploitative and instrumentally regimented care, oriented by a one-way anthropocentric temporality.

This direction could be troubled by moves perceptible in the way the soil sciences are reconceiving how they see soil as a natural body, with important consequences about how to care for it. We can see changes supported by a notion that soils are of more "use" than agricultural production. An emphasis on the multiplication of "soil functions" (Bouma 2009) means that they are valued for other purposes than agriculture, or building. This points at a diversification of the applications of soil sciences as soils become providers of a range of "ecosystem services"—for example, including social, aesthetic, and spiritual value—beyond commercial agricultural needs (Robinson et al. 2014). The ecosystem-services approach looks at the elements involved in an ecological setting or landscape from the perspective of what they offer to humans beyond purely economic value and tries to calculate other sources of value—not necessarily to "price" them, a distinction important to many advocates of this approach. This is a significant move for human–soil relations with a transformative potential that shouldn't be underestimated. Yet this notion has its limitations to transform the dominant affective ecologies of human–soil relations and not merely because it is restricted to a calculative vision of relationalities. Even if we accepted staying within a logic of valuation and service provision, at the very least a notion of ecosystem services should also calculate those provided by humans to sustain a particular ecology and the nonhuman community. The notion of ecosystem services, while representing an important attempt from inside Capitalo-centered societies to shift the parameters of a purely economistic valuation of nature for production, is not enough to bring us closer to a relation of care that disrupts the notion of other than humans as "resources" and the sterile binary of utilitarian versus altruistic relations with other than humans. A notion of care, Sue Jackson and Lisa Palmer argue, could disrupt this logic and improve the way ecosystem services are conceptualized:

> If we extend the concept of relatedness from humanity to all existence and foster an ethic of care which recognizes the agency of all "others," be it other people or other nature, and the specific cultivation of these relations by humans, we avert the broadening of a schism between nature and culture— the schism that in the ecosystem service framework construes nature as provider/producer and human as consumer. (Jackson and Palmer 2015)

Thinking with a feminist politics of care that remembers the contested exploitations involved in the type of service work that care is often made to be, we can also interrogate the connotations involved in the notion of "service" itself. While service could seem to lead us beyond a logic of exchange—doesn't service also refer to what we do for altruistic purposes or sense of duty?—in strongly stratified societies it is marked by a history of serfdom. Struggles around the relegation of domestic care to women's work showed how the point is not only to make this "service" more valuable or recognized but also to question the very division of labor that underpins it. A feminist approach to more than human care would at the very least open a speculative interrogation: *Cui bono?* (Star 1995) service *for whom?* as a question that reveals the limitations of a service approach to transform human–soil relations while it remains based on conceiving naturecultural entities as resources *for* human consumption, thus interrogating an understanding of soils that posits them as either functions or services to "human well-being" (Millennium Ecosystem Assessment 2005). An interrogation of both the productionist and service logic can learn from ecofeminist critiques about the intrumentalization, degradation, and evacuation of more than human agency (see, e.g., Plumwood 2001; Bastian 2009) and the connection of these ecologically oppressive logics to gender and racialized binaries with their classic segregation of life domains (Mellor 1997). Thinking with care invites us to question unilateral relationalities and exclusionary bifurcations of living, doings, and agencies. It brings us to thinking from the perspective of the maintenance of a many-sided web of relations involved in the very possibility of ecosystem services rather than only of benefits to humans. Coming back to re-articulating relations of care and temporality, I engage below a conception of soil "as living" that can further question its persistent status as *serving* for input for crop production or other human needs. A more soil-attentive mode of care might also reveal other ways of experiencing time at the heart of productionist relations, while, as Haraway would put it, "staying with the trouble" of humans' relation to soil as an essential resource for survival.

The Living Soil: Becoming in the Foodweb

As part of the ecological turn, soil ecology research has become more important at the heart of the soil sciences, concentrating on *relations* between biophysical, organic, and animal entities and processes (Lavelle 2000; Lavelle and Spain 2003). Moreover, a number of accounts of the discipline's development in the past ten years connect the growing significance of the ecological perspective with the moving of biology to the center of a field traditionally dominated by physics and chemistry. In this context, it is remarkable how a notion of "living soil"—once mostly associated with organic and radical visions of agriculture (Balfour 1943)—is now mainstream. This does not mean that soil science traditionally conceived of soils as inert matter. Even conceptions of soil as reservoirs of crop nutrition focus on lively physicochemical processes and interactions. Also, soil microbiology has been a crucial part of soil science since its early beginnings as well as is important precursor work on soil biology (such as Charles Darwin's work on earthworms). This does not mean either that biology and ecology support environmentalism per se or that other disciplinary orientations in soil science must now be connected to biology. The noticeable changing trend is the increased significance of "biota," from microbial to invertebrate fauna and, of course, plants, roots, and fungi, in the very definition of soil. That this has not been an obvious move is attested by ecologists who claim for a change in soil's definitions:

> Are living organisms part of soil? We would include the phrase "with its living organisms" in the general definition of soil. Thus, from our viewpoint soil is alive and is composed of living and nonliving components having many interactions. . . . When we view the soil system as an environment for organisms, we must remember *that the biota have been involved in its creation, as well as adapting to life within it.* (Coleman, Crossley, and Hendrix 2004, xvi, emphasis added)

In this conception, soil is not just a habitat or medium for plants and organisms; nor is it just decomposed material, the organic and mineral end product of organism activity. Organisms *are* soil. A lively soil can only exist with and through a multispecies community of biota that *makes* it, that contributes to its creation.

One of the most significant aspects of these changes in conceptions of soil is a growing interest in investigating biodiversity as a factor of soil fertility and system stability (Wardle 2002, 238, 234). This goes beyond biological interest; for instance, the recognition of the importance of large pores in soil structures gives a central place to increased research on soil fauna such as earthworms, which some have named the "soil engineers" (Lavelle 2000). In the words of a soil physicist: "As the appreciation of ecological relationships in soil science developed after the 1970s, studies on the role of soil animals in the decomposition process and in soil fertility have been more common" (Warkentin 1994, 8). More research focuses on the loss of soil biodiversity after alterations (van Leeuwen et al. 2011) and on the ecological significance of soil health for nonsoil species (Wardle 2002). A number of soil scientists are now engaged in drawing attention to biodiversity in soils as part of educational campaigns and soil fertility projects worldwide.[8] Soils have become a matter of concern and care not just for what they provide for humans but for ensuring the subsistence of soil communities more broadly.

These developments are not disconnected from worries about the capacities of soil to continue to provide services (a range of calculations are deployed to value the services of biota) or from a notion that accounts for soil fertility according to its ability to provide yield. Production continues to be a concern as the "loss of organic matter, diminishment or disappearance of groups of the soil biota and the accompanying decline in soil physical and chemical properties" are identified as important causes of "yield declines under long-term cultivation" (Swift 2001, xx). However, these approaches bring significant hesitations at the heart of a conception of soils as physicochemical input compounds. Soils as living, for instance, create other questions about effects of human interventions to technologically enhance impoverished soils, however well intentioned. For example, agrochemical inputs can benefit crop yield, but soil communities can face long-term destabilization or destruction, making soils and growers dependent on fertilizers. Also, the protection of soil structures connects to a generalized reevaluation of tillage in agriculture and other technologies that alter and destroy fragile and complex soil structures.[9] In sum, exploiting soil species for production threatens to destroy the living agents of this very productivity (Tsiafouli et al. 2014). Once again, reconceptualizations of soil as living emphasize how productionist practices ignore the complex diversity of soil-renewal processes in favor of linear temporalities aimed at speeding up abundant output.

It is the nature of soil itself and ways to care for it that are at stake in these moves. Attention to soils as a living multispecies world involve changes in the ways humans maintain, care, and foster this liveliness (Puig de la Bellacasa 2014a). So how does this affect temporal involvements in caring for the soil as a multispecies world? I approach these through the example of the "foodweb," an ecological model of soil life that, having become popular in alternative growers' movements, thrives at the boundaries of soil science.

Foodweb models are not new, but they became increasingly prominent in soil ecology after the 1990s (Pimm, Lawton, and Cohen 1991). Foodweb models are valuable for scientists to describe the incredibly complex interactions between species that allow the circulation of nutrients and energy. They follow predation and eating patterns as well as energy use and processing. Soil foodweb species can include algae, bacteria, fungi, protozoa, nematodes, arthropods, earthworms, larger animals such as rabbits, and, of course, plants. They describe not only how species feed on each other but how one species' waste becomes another one's food (Coleman, Odum, and Crossley 1992; Wardle 1999; Ingham 2004). Foodweb conceptions of soil question the use of artificial fertilizers, pesticides, and intensified agricultural models more generally. This is because their weblike, interdependent configuration means that altering or removing any one element can destroy them. Often conceptualized as soil "communities" even as they are based on "trophic" relations—who eats whom—foodweb models emphasize a living world below, teeming with life and yet always fragile. Soil ecology is, of course, not a unified domain and, while rich in holistic models of life cycles, it is also rich in reductionisms. If I am lured by moves that see soil as a multispecies world, it is for how they could affect not only the nature of soil itself but also the ways humans maintain, repair, and foster soil's liveliness—that is, the agencies involved in more than human webs of care.

Interdependent models such as the foodweb disturb the unidirectionality of care conceived within the linear timescapes of productionist time traditionally centered in human-crop care relations. Relational approaches to the cycles of soil life in themselves can be read as disruptions to productionist linear time, simply because ecological relations require taking a diversity of timescales into account (Bird Rose 2012). Yet foodweb models also affect relations to soil for how they turn humans into full participant "members" of the soil community rather than merely consumers of its produce or beneficiaries of its services. It is the emphasis on the interdependency of soil communities that is appealing for exploring more than human care as an immanent obligation that passes through doings and agencies involved in the necessary maintaining, continuing, and repairing of flourishing living webs. Remembering discussions in previous chapters around the nonreciprocal qualities of care, we see that while care often is represented as one-to-one practice between "a carer" and "a cared for," it is rare that a carer gets back the care that she gives from the same person who she cares for. Carers are themselves often cared for by someone else. Reciprocity of care is asymmetric and multilateral, collectively shared. A caring conception of soil emphasizes this embeddedness in relations of interdependency. Caring for soil communities involves making a speculative effort toward the acknowledgment that the (human) carer also depends on soil's capacity to "take care" of a number of processes that are vital to more than her existence. Thinking multispecies models such as foodwebs through care involves looking at the dependency of the (human) carer not so much from soil's produce or "service" but from an inherent relationality. This is emphasized by how the capacities of soil in foodwebs refer to a multilateral relational arrangement in which food, energy, and waste circulate in nonreciprocal exchanges. Foodwebs are therefore a good example to think about the vibrant ethicality in webs of interdependency, the a-subjective but necessary ethos of care circulating through these agencies that are taking care of one another's needs in more than human relations.

A care approach needs to look not only at how soils and other resources produce output or provide services to humans but also at how humans are specifically obliged, how they are providing. The capacity of exhausted global soils to sustain these webs of relations has become more dependent on the care humans put in them. In resonance with Anthropocenic narratives that acknowledge the impact of situated human actions on the making of earth, what the above conception might require is not only for organisms but also for humans to be included more decisively *in* the concept of soil. Here, in turn, changing ways in soil care would affect soil ontology. Coming back to the redefinition of soil as living (Coleman et al. 2004), we could include a rephrasing such as: "When we view the soil system as an environment for humans, we must remember *that humans have been involved in its creation, as well as adapting to life within it.*"

Though scientists have long spoken of "soil communities" to refer to the organisms involved in soil's ecology, the idea that humans are part of soil communities is not a prevailing one in the scientific literature. Scientific illustrations of the soil foodweb rarely represent humans as part of this relational web—for example, as producers of "organic waste" and beneficiaries of the output of plants. This could be connected to the traditional role given to the anthropogenic element in soil scientific literature, where it is generally considered as one "element" of soil ecosystems and formation processes that "lies apart" because of the higher impact of its activities in a shorter amount of time than other organisms. The "human" mostly features as an unbalanced irruption in soil's ecological cycles—or a victim in the case of soil pollution—rather than as a "member" of a soil community (Hillel 2004). Notions of humans as members, or even of humans *being soil*, thrive outside science, however—including in how scientists speak of soil (and land) beyond their "official" institutional work (Hole 1988; Warkentin 2006). It could be argued that alternative affective ecologies with soil become obscured within science. But in the spirit of staging matters of fact, scientific things, as matters of care, it seems to be a more fertile option to attempt an articulation of different horizons of practice and modes of relating to soil through their potential to transform human–soil relations. Connections with "nonscientific" ways of knowing soil, whose relevance is sometimes also mentioned by scientists (Tomich et al. 2011), could become even more important in the light of an argument for a shift in soil models from considering soil as a "natural body" to soil as a "human-natural" body (Richter and Yaalon 2012) and for the introduction of new approaches such as "anthropedology" that broaden soil science's approach to human–soil relations (Richter et al. 2011).

Now, like all Anthropocenic narratives, these ideas would require nuancing which Anthropos is being spoken for, asking questions such as: If the marks on Earth that are to be accounted for are those that dramatically altered the geological makeup of the planet since the industrial age or atomic essays, shouldn't we, as Jason Moore argues, rather declare a Capitalocene? Or, should we, as Chris Cuomo has called for, reject this recentering of the notion of Anthropos altogether for its masking of capitalist and colonial dominations.[10] Or, couldn't we propose—questioning the tendency of Anthropocenic thinking to further evacuate agency from the other than human world and to reinstate Man as the center of creation—populate our speculative imagination with visions of more than human coexistent epochs that amplify the proliferation of symbiotic processes with multifarious nonhuman agencies such as Haraway invites us to do with a *Chthulucene* (Haraway 2015). All these doubts contribute to complicate the narratives of the agential ethicalities at stake in reinstating humans in the concept of soil. Desituated storylines of Anthropos-centered relations need to be challenged if are we to offer situated humans a place within, rather than above, other earth creatures, in acknowledgment of specific modes of agency: a vital task for environmental thought and practice, across the social sciences and humanities, but also for exceeding collective imaginations.

The exploration of decentered ethicalites of care via foodweb visions of human–soil relations can be nourished by such collective imaginations to contribute a displacing of human agencies without diluting situated obligations. Eliciting articulations of the sciences with other domains of practices, even subtle, is important here. Obviously, my reading of foodweb models goes beyond its explanatory potential to alter scientific conceptions of soil. Speculative thinking is professedly excluded from scientific concerns maybe even more than political stances. But when understood as part of a naturecultural transformation in human–soil relations of care, the foodweb is not just a scientific model. One could say that successful scientific models owe part of their power to their figurative potential. Beyond science, the foodweb is a charged figuration of soil relations, which I read here as going in the sense of restoring what Thompson calls the "spirit of the soil," by which he points at an understanding of human activity as part of the life of the earth and "the spirit of raising food and eating it as an act of communion with some larger whole" (Thompson 1995, 18).

The search for glimpses of a transformative ethos in human–soil relations moves us beyond science and its applications to the articulations of alternative affective ecologies and technoscientific imaginaries to which science participates but not necessarily drives. The soil foodweb model is interesting in this regard because it has become, beyond science, a symbol of alternative ecological involvement—particularly in ecological movements where alternative visions of soil practice are being developed, such as agroecology, permaculture, and other radical approaches to agricultural practice. It is in these conceptions that transformative trends in soil relationalities can be read most visibly for how they foster a different relation of care, one susceptible to alter the linear nature of future-oriented technoscientific, productionist extraction in anthropocentric timescapes.

Making Time for Soil (Care)

Beyond science, foodweb models and scientific ideas of soil as living are explicitly made to speak for alternative soil-care and human–soil relations, with implications for the dominant productionist futurity. I first learned about them by following the work of Elaine Ingham, a soil scientist specializing in foodwebs, who is influential in the teachings on composting at EAT trainings in permaculture. Originally a microbiologist working in the field of soil ecology, Ingham's work on foodwebs is cited in scientific publications until the early 2000s, but then becomes mostly visible through her work beyond academia. She left Oregon State University to lead her foodweb-based soil-testing company and subsequently became Chief Scientist of the Rodale Institute, which promotes organic agriculture.[11] Ingham also directs her own Sustainable Studies Institute and has an impressive online presence as a celebrated adviser of alternative soil care.

Among her many interventions, I was enticed by a series of online lectures in which Ingham popularizes a "biological" notion of soil among practitioners: soil is not "dirt." Dirt is soil without life, she affirms. Here she introduces the basics of microbiology to inform accessible soil-sampling techniques and subsequent soil testing, including how to choose a secondhand microscope and set it up to sample soil. In a clip, "How to Take a Soil Sample: Introduction to Soil Microbiology,"[12] Ingham walks out in a patch of unattractive grass explaining how to sample soil to examine "the biology present." She then presents in an amusing tone the instrument she is going to use: "this *really expensive* high-tech piece of equipment called an apple corer." She shows how to introduce it into the soil, in round movements, "just like going through an apple," to get to the decomposed part under plants and roots. The aim is to get at "the biology" in soil. Noting the unhealthy look of a small patch of grass she is touching, probably revealing "disease problems," she affirms that the life in this soil needs some "biological" help: "But *what* biology do we need?" she asks. "That is what we are really interested in figuring out this way." In the following clip, "How to Prepare and View a Soil Sample under the Microscope," Ingham has gone back into the house, where she had set up a microscope and other instruments, and explains how to give an estimate of the number of bacteria present in the soil. This methodology to assess soil health is based on an estimated count of microorganisms and aims at detecting the needs of soil in order to feed it with appropriately balanced organic material, such as compost and compost teas produced on-site (Ingham 2000).

The Queen of Compost is extensively named in soil lovers' worlds as having produced science-based techniques that improve growers' practices. Yet Ingham also has an explicit political ambition for her educational quest—to liberate farmers from industrial fertilizers. "Jump off the chemical wagon!" she calls in a video advertising her courses. Her trajectory reflects contemporary renegotiations of scientific spaces between academia, business, and public engagement—soil-testing business or advising farmers are not atypical career paths for soil scientists. But throughout her work of propagation of the soil foodweb model, there is a sense that she is making soil science available in a classic *for the people* way. As an activist with scientist credentials (and vice versa), Ingham also communicates with a world of amateur soil scientists that are only starting to join the institutionalized forms akin to more established modes of "citizen science" projects (Charvolin, Micoud, and Nyhar 2007).

This is scientific work, but it is displaced, situated, implicated, involved, and "distributed" in technoscience (Papadopoulos 2014a; 2017). This activist science is not in the purist position of an outsider. Ingham's vision mobilizes "science-informed" soil practice as a promise of future output: effortless, chemical-free, and abundant yield (Ingham 1999). It could be said that the message is compelling because it still speaks to the production ethos as a shared hope of growers to benefit from abundant produce from a fertile soil. Yet here production is harnessed by good care rather than the contrary; and good care is tied to knowing and appreciating soil life. These practices speak of intensification, intensification not so much of production but rather of *involvement* with soils. These modes of soil care involve practitioners with the agencies and mediations that make the soil community work well, that is, capable of taking care of biological "functions" in ways that would be made invisible by off-site testing practices. Ingham is inviting soil practitioners to immerse themselves in the soil and develop their "feeling for the soil," to paraphrase Evelyn Fox Keller (1984; see also Myers 2008).

Affective involvement with soils is all but alien to farming practices (Münster, forthcoming), as Guy Watson, a UK farmer and founder of Riverford Organics, puts it: "Some farmers speak of intense affective relations with soils of how they feel acutely protective of their soil, treating it with the commitment, concern and empathy normally reserved for close family members. I have seen organic farmers sniffing and even tasting their soil, and disrobing its virtues with familiarity and affection." Here again, knowing about by seeing/touching *the life* in the soil acts as a powerful signifier of a greater proximity and care with its dark alterity:

> A handful of healthy soil can contain millions of life forms from tens of thousands of different species. . . . Pesticides, fertilisers, animal wormers . . . can all drastically reduce these populations, not by just a few percent but by 10 or even 100 percent. Imagine the outcry from WWF if anyone could see the carnage. . . . So if you can't see the fungi, bacteria and invertebrates and you don't feel inclined or qualified to taste your soil, how do you know it is healthy? (Watson and Baxter 2008, 14)

The idea that affective involvement can be provoked by "seeing" soils as living is not alien to scientific and academic circles. A scientist involved in the Global Soil Biodiversity project argues that showing images of the organisms to farmers and growers opens the soil "black box" and invites us to "identify . . . with soil fauna."[13] But testing soil as "tasting soil," treating soil as family, notions of immersing into soil and comingling with its substance, speak of sensorial involvements with a soil that is not conceived as separate. And yet these intimate affections bring us back to the haptic engagements and immanent obligations of caring. We could read as common in these instances a desire to reduce distance and separation in visual-haptic yearnings of closeness (discussed in chapter 4). The question is, then, whether it is by seeing, touching, or tasting, what is this feeling-for the soil standing for? What does it mean to "identify" with soil fauna? In discussing haptic involvements, I emphasized an "unknowability of the other," and the caution remains for precise immersed microscopic imagery as well as idealizing direct touching of the soil. What/how do I see? What/how do I touch? Closeness is not necessarily caring more or better, so what is Ingham's practice calling for in her way to account for previously ignored or neglected beings?

The importance of asking such questions reminds me of Astrid Schrader's insights on care, as she wonders: "How do we begin to care about others of whose existence we might not even have been previously aware, let alone teach others to care?" (Schrader 2015, 3). Explicitly challenging identification—"We simply cannot find ourselves in these creatures"—she asks crucial questions, such as: "Would it be possible to begin to care without an a priori identification or categorization of an object of care? Can we conceive of a less anthropocentric notion of care that is attentive to inde-terminacies in its practices?" (Raffles 2010, 44, cited in Schrader 2015, 15).[14] Schrader explores the affectivity of care as an in-between "abyssal intimacy" that leaves the subject of care indeterminate, the act of care undecided, and thus reconfigures time and space in relations to the other and scientific knowledge production. "Abyssal intimacy does not require recognition, but describes a creative engagement that relies on the withdrawal of the self, a passivity that enables an active listening, an opening to surprises" (Schrader 2015, 9). These are crucial questions for learning to perceive previously neglected soil lives. A form of passivity—as withdrawal of self but also of identified outcomes—seems vital in a conception of the relational web of care that troubles productionist relations, in ways kin to the indeterminacy of being "drawn to," in Povinelli's immanent conception of obligation.

Schrader is specifically discussing the affectivity of "caring about" those we have not previously cared for, a relation that does not necessarily conclude in an act of direct care. This emphasis on the affective side of care allows me to bring forth a contrast with Ingham's activist "educational" projects that I believe shifts the position of ethical questioning. Her efforts are not directed to learning to care about, to be affected by, something that was not cared for before. Ingham speaks to an already affected constituency. Recognizing the worth of bacteria and other microbial nonhumans points to an affective engagement in which knowing-caring is intrinsically soil practice, whether in farming or in science. Thinking with doing/work-affect-ethics/politics allows us to emphasize this embeddedness in everyday practice. And therefore the intensification of proximity does not happen over an abyss; it is about learning to care differently within existing modes of taking care, displacing affectivities as the doings move. The reason Ingham suggests that practitioners engage with DIY testing is to create a sense of a communal character of these doings, the living web that provides. In other words, the wonder here is not about ethical recognition of beings we feel as radically other to humans. As Papadopoulos reminds us, alterontologies happen as particular humans create more than human rearrangements with particular nonhumans. The appeal is transformative because it connects to an already recognized necessity of taking care of soils while gradually displacing this ethos with other affective and ethical sensitivities. Inviting practitioners to make time to see bacteria or other microscopic beings—count them, feel them, learn to feed them well—engages curiosity toward a web of doings, obligations, and asymmetrical reciprocities that practitioners can easily conceive: the soil you depend on depends on those who depend on you. This is what fertilizers and pesticides can destroy. The recognition here is neither symmetrical nor identificatory. The proximity is based on an everyday relation rather than fascination or aversion. The point is not so much to translate care into acting—acting is already there in practices of maintaining soils—or to care about something that was previously unknown, but to alter existing relations of taking care through alternative modes of affectivity. What this brings forth is that ethical recognition of other than humans might not always prompt questions about other than human alterities but rather about modest changes in our ethos of living with many others, by creating mundane paths for our doings that acknowledge how we are already ordinary everyday companions. The foodweb as a figure of alternative caring relations with soil works that way: it offers new obligations within existing ones, immanent obligations that could unfold uneventfully, ordinarily.

Coming back to the articulations of care and temporality, I turn to how these reorientations of soil care, as they subordinate production to immersed relation, trouble the linearity of productionist futurity and extraction from the soil dominating in contemporary technoscience. With regard to epochal progressive futurity, and amid calls for urgent and global responses to food insecurity, these small-scale reorientations of growers' skills are bound to appear as insignificant attempts to "turn back the clocks" to preindustrial practice. Similarly, from the perspective of "bright and shiny" conceptions of innovation, tasks such as "counting bacteria" to test soil health recall school science projects. Ingham's work projects a sense of outdatedness, exaggerated by the use of tools like an apple corer and secondhand microscope. From the perspective of the embedded temporality of practice, one can wonder why a busy farmer or gardener preoccupied with output constraints would *make time* for these slow, labor-intensive tasks, instead of putting soil into an envelope and sending it to a soil-testing company. In fact, what we see here is akin to what Patrick Bresnihan elicits in his ethnography of fishermen's "commoning" practices. Bresnihan exposes modes of management of fish stocks that are at odds, though cohabiting, with the standard management of sustainability. Here, alternative engagements with time are at stake that not only evoke a different mode of production, but a different mode of life, including a different relationship to work. This temporal relation is not focused on "efficiency," and because of that it seems inconceivable from the perspective of the "rational calculations of a liberal subject plotting his activities along a more or less individualized and linear trajectory," that is, the perspective of "management . . . where the future is organized toward a specific, technically defined goal of biological sustainability" (Bresnihan 2016). In a similar way, the embodied experience of time in making time for soils alters linear productive practice in ways that remain irrelevant, but also potentially disruptive, to the perspective of the trajectories of productive futurity in technoscience.

To further illustrate this I draw upon a discussion of "time niches" extracted from an influential manual of permaculture, a movement that counts with numerous foodweb proponents. The author, Bill Mollison, speaks of an embodied immersion in ecological cycles that involves a long period of "thoughtful and protracted observation" before acting on the land and its processes. This principle, known as "TAPO," is a rule of technical design and an ethical principle in training in permaculture practice (Ghelfi 2015). The point of immersed observation is to take the time to "experience" the specific "schedules" happening within the arrangement of life cycles (involving species, climate, localized interactions, etc.) that constitute temporal niches in a particular ecology (Mollison 1988, 28). The imperative of observation is an ongoing one because each cycle is an "event": "diet, choice, selection, season, weather, digestion, and regeneration differ each time [the cycle] happens" (23). It is in such variation that the possibilities for diversity thrive. Soil-care practitioners often speak about similar kinds of immersion in the repetitions of cycles of soil life, by which they learn the needs of the landscape and by which a particular ecological environment also "learns" and adapts to human practice. In this conception, TAPO is about learning to work with these cycles as a mode of relational involvement required by appropriate ecological care. TAPO is an ethos that contributes to the cocreation of a particular ecology and the mutual multilateral obligations and interdependent doings it entails.

Soil ecologists have long been aware of cycles of interdependent growth and decay in the living soil that articulate multiple temporalities. The temporal immersion of TAPO is specifically oriented to rethinking human ecological practice in its material obligations with ethical and affective dimensions, that is, care. TAPO requires making time for the times of the soil and, I argue, can be read as a form of cultivating "care time." First, the repetitive character of ongoing observation of soil cycles enables care. Care work becomes better when it is done *again*, creating the specificity of a relation through intensified involvement and knowledge. It requires attention and fine-tuning to the temporal rhythms of an "other" and to the specific relations that are being woven together.[15] Second, TAPO's temporal immersion involves human practice in an interdependent, yet diverse, web. Temporal diversity, rather than immediate connection (to nature) or mere control of other rhythms, needs to persist in these tunings and readjustments. One form of care does not necessarily work in a different arrangement, nor do different temporalities cohabit in harmony. Different types of soil will need different care and members of the foodweb are often read as competitors.

TAPO is specifically immersed, or embodied, observation of a specific ecological community that obliges; observing cycles and processes here is not only about becoming aware of them but about a requirement to tune in to these rhythms. In terms of human–soil relations more generally, practitioners are not so much "in charge" of ecological management and food production than they are attentive members of a specific ecological, soil foodweb community. This ethos disrupts humans' location as outsider observers or central beneficiaries of objectified services: even if it strongly relies on the role played by humans in landscapes that they are part of, humans are not the end destination of the processes that human–soil ecosystems take care of. The *as well as possibleness* of the webs of care is dependent on a potentially immeasurable mesh of interdependent agencies. Within these conceptions, to properly care for the soil, humans cannot be only producers or consumers in the community of soil-making organisms but must work, and be, *in* the relation with soil as a significant living world. Participants in a foodweb somehow embody the time of the cycle by eating or becoming food for other participants in the death and decay cycle.[16] There are affinities here with the intimate relation with soil cultivated by farmers described by Kristina Lyons's immersed ethnographies of human-soil relations in the Amazonian plains. She emphasizes modes of relation that do not set soil apart as a separate entity from plants or humans. All of these beings play a part in embodied and sensorial involvements: humans become "one among many actors who labor in the act of living and struggling together," but they also cultivate a specific obligation: to have "eyes for her"—for the *selva* (Lyons 2014). Another obligation created in this relational web is what Lyons calls "decomposition as life politics" through the circulation of waste-food/death-life/decay-regeneration (Lyons 2016). Immersion in a foodweb *as life politics* creates specific practical eco-ethical obligations, such as the cyclic return of organic waste (i.e., through composting, as I discussed in the previous chapter). One care task here is, as gardeners like to put it, to *grow soil* (Bial 2000) by "returning the surplus" in order to continue to make soil as much as we consume (from) it—an enactment of interdependent care.

Ecological models such as the foodweb are not only about knowing the soil better so that we can extract more efficiently from it but about another way of relating, about the thicker, haptic, involvements and embodied traffics in a more than human community of soil makers. Focusing on these forms of immersed ecological care, we can sense changes in human–soil relations based on material, ethical, and affective ecologies: how to qualify the affective ecologies involved in these transforming practices of soil care and ways of making time for ecological relationalities. Looking at imperceptible doings of care thickens the dominant timescape with a range of relational rearrangements. In these relations of care the present is dense, thickened with a multiplicity of entangled and *involved* timelines rather than compressed and subordinated to the linear achievement of future output. Across this book I have often used the notion of involvement as a synonym of engagement, of committed relations and politics. Involvement acquires deeper temporal meaning thinking with Carla Hustak and Natasha Myers (2012; 2013), who speak of "involutionary momentum" to name the occasion for a new relational arrangement between species. The involutionary has a nonlinear temporal quality—not an evolutionary move, not a coevolution, but an *intensification* of involvements and mutual co-envelopments. Shared experiences and temporal tunings of relations of care with the living soil could hopefully be involutionary, intensifying attentiveness within already existing relations of interdependency and mutual involvement, rather than setting ethical expectations on a teleological event that would shift species activity.

The Disruptive Pace of Care

Reenacting care as a disruptive intervention is an involvement in the mattering of worlds. I have found temporal matters in human–soil relations to be an illuminative terrain to engage with the complex ambivalences of the living webs of care in the more than human worlds of technoscience and naturecutures. I have been drawn to think of care-time as that which cohabits but remains imperceptible from the perspective of anticipatory-futuristic science. It is by engaging with care time as disruptive that this

book ends, hopefully opening possibilities for thinking with care in worlds
that disregard its transformative potential.

I have engaged with care in generic terms as a doing that is always
specific (one form of care is not necessarily transposable somewhere else).
Feminist materialist discussions of the experience of care as socially embodied agencies have been crucial to this understanding. People become
"obliged" to care in actual practices and relational arrangements, in messy
material constraints rather than through moral dispositions. The open
question *How to care?* (which I asked in chapter 1 as a premise for specific discussions of care) grounds care ethics in situation. This wondering
remains a critically troubling question that entails unpacking what is actually done under the name of care, whatever good the intentions. Care is
not only political, messy, and dirty; it is a trap for many and not less in
technoscience. But asking *how to care* is an open wonder about the ethico-
political significance of doings of care as immanent obligation. So while a
critical stance can bring attention to such matters as who cares for whom,
to what forms of care are prioritized at the expenses of others, a politics of
speculative thinking also is a commitment to seek what other worlds could
be in the making through caring while staying with the trouble of our
own complicities and implications. Feminist visions of care emphasize the
ethico-political significance of doings of care that make the substrate of
everyday life, not as a *separate* cozy realm where "nice" relations can thrive.
Noninnocent thinking resides in the inevitable entanglement between the
critical and speculative stance: there is no position from where to claim
having the answer of what as well as possible care means, or not. And that
there is not such an outside position also means our involvements have
effects. Staying within this unstable stance, this book has proposed, however, that paying attention to the worlds of care, holding together a plural-
ity of ontological meanings—doing/work, affect/feel, and ethics/politics—
is a disrupting and hopeful way to be involved with the lively and consequential ethicalities that are being drastically reconfigured in the more
than human worlds of technoscience.

And so looking at human–soil relations through the articulations of
temporality and care both critically exposes the prioritization of anthropocentric technoscientific futurity and makes visible *coexisting* alternative

timescapes, possibly enriching temporal imaginings. Pervaded by soil's ancestral status as a resource, as a crop receptacle, and by a temporality subjected to increase yield, productionist relations to soil remain predominant, and so it is likely that agricultural intensification and increases in chemical fertilization will be immediate responses by agribusiness and policymakers to future food-security alarms. I have stressed questionings to the dominant treatment of soils from within: from tensions in soil science around the imperative of progress, to moves away from productionism toward conceptions of soil as living, and correlated practices of involvement with soil as a foodweb that humans are part of. These immersions in soil times do not exist in an unpolluted temporality that would sit as alternatives outside the current crisis. Experiences of intensified care time could disrupt the futuristic drive, but they are not disentangled from technoscientific time. I learned to appreciate this through Chris Kortright's ethnographic work on GM rice research. Kortright shows how forms of creative and intense caring labor exist in scientific practices working in the development of genetically modified rice plants destined to serve a second green, genetically modified, revolution (Kortright 2013). While this is another context that makes the case for an enlargement of frames— *for what is care being done?*—it also shows that more than human care does not exist in an alternative timespace altogether *because* being "cared" for in one way or another is a condition for all beings living in the currently uneven, asymmetrical web of more than human interdependencies. We could even argue following Dimitris Papadopoulos that the practices I have approached in these two last chapters are also technoscientific, making alternative ontologies from within this timescape (Papadopoulos 2014b). And so the question of what worlds will (our) care become enrolled in sustaining becomes even more acute.

Care is a necessary everyday doing, and as such it carries a compelling character that can become the moralistic justification under which regimes of power and control circulate and justify. Feminist work has analyzed and contested the ways in which the everyday needs of care became a burden largely assumed by women. As noted before, this critique grounds the ethico-political relevance of eco-ethical questions such as: Who provides the ecosystem "service" and for whom? And yet critical attention to discrepancies in the everyday doings of care also makes visible alternative affective ecologies and embodied traffics, disadjusted timescales, and a great deal of oppositional work at stake in attempting to maintaining as well as possible interdependent relationalities in worlds that privilege autonomy and self-sufficiency. Emphasizing these alternatives, thinking with care intervenes in the mattering of worlds. In this direction I delineate below tensions and transformations perceived in alternative timescapes of soil care that could be reworking predominant notions of futurity and innovation from within. I am reading ways of making time for soil as "care time" that is made irrelevant from the perspective of the progress-oriented, productionist, restless futurity that I have identified previously as the predominant technoscientific timescape at epochal, practical, and embodied levels.

Starting with embodied time: a focus on care elicits material and affective involvements at stake in maintaining and fostering interdependent human–soil relations. These include adjustments according to cycles, articulating future and past in a presently embedded time, as well as different ecological timescales. Feminist anthropologies of caring practices can support this observation, for they expose labors of everyday mundane maintenance, repetitive work, requiring regularity and task reiteration (Mol 2008; Mol, Moser, and Pols 2010; Singleton and Law 2013). Anybody who has been involved in caring for children, pets, elderly kin, an allotment, cells in a petri dish, knows that the work of care takes time and involves *making time* of an unexceptional particular kind. It requires having to deal with necessary material doings of maintenance that absorb time but ground the everyday possibility of living as well as possible—cleaning up vomit or digging ditches. Affectively, this aspect of care time can be enjoyable but also very tiresome, involving a lot of hovering and adjusting to the temporal exigencies of the cared for. Ethically and politically this work remains neglected, receiving among the lowest wages, even within sociotechnical regimes that put carers under high moral pressure because of the economic importance of their work.

Care time is not about harmonizing dislocated time. Following the intrinsically relational meaning of care approached in chapter 3, as concomitant to complex heterogeneities, care cannot be holistic in the sense of aiming to recover "a sense of oneness." It goes in the opposite direction as it requires "an understanding of *the real difficulties in the way of fostering the growth of concrete, multifaceted, caring relations* among individuals, societies and the nonhuman beings and systems among whom they live" (King 1991, 80). Moreover, temporal diversity is crucial in tunings and readjustments of intensified involvements because one form of care does not necessarily work in a different arrangement, and will need to be readjusted as a relation evolves (e.g., different types of soil will need different care, even the same soil at different times of the year). Given the recalcitrant diversity of soils, unexpectedness and indeterminacy are part of care work because specific relationalities are at stake and therefore need a fair amount of "tinkering" (Mol, Moser, and Pols 2010). But it is also true that specific care becomes better when it is done again, in the particularities of a knowing relation that thickens as it goes, as it *involves*. I have noted earlier how future, urgent, speedy temporality suspends and compresses the present. It could be thought that care time suspends the future and distends the present, thickening it with myriad multilateral demands. It would be wrong to purify the time of care from other timescales—for instance, it is my yearning to see my child grow into the future and these thoughts are also affected by a sense of mortality, fears, and possible anxieties, and of course by the weight of passing-presents and lessons of caring that we are repeating, reenacting. Care time is not a get-it-while-you-can *now*, which ignores the future and obliterates the past. But even when one cares for the dying, with hope and anxious anticipation, even when care is compelled by urgency to enjoy the fleeting present, charged by past regrets and joys and the weight of accumulated experiences, a certain suspension of feelings of emergency, fear, and future projections—and weighty pasts—is required to focus on caring attention. In particular with regard to anxious futurity, feelings of emergency and fear, as well as temporal projections, need often to be set aside in order to focus and get on with the tasks necessary to everyday caring maintenance. Without this mode of attention, care would be an impossible charge, always at the edge of a break.

Coming back to soil care, while the probability and repetition of ecological cycles coexists with uncertainty and restless anxiety about future unexpected events (one only needs to think of weather, pests, disaster, climate-change anxieties), some expected repetitiveness—grounded on trust and reliance on the continuity of relative life processes—is part of ecological relations of care. Taking care, even of the unexpected, remains an unavoidable immanent obligation for those living in more than human webs. Equally, relations of care are made more difficult when we are under the pressures of managerial and output-oriented time constrains. Unpaid and paid carers often ask not so much to be paid, or to be paid more, but to be allowed more time to care well (Ehrenstein 2006). Caring attention needs a certain abstraction from the discontinuity of time and the compression of the present that marks anticipatory preemptive technoscience. The risk-taking ethos of promissory and anxious futuristic technoscience obscures the quality and persistence of these everyday doings. This is not to promote a conservative notion of time; drawing attention to this timescale is rather a refusal of the binary that dismisses it with regard not only to linear, managerially predictable, conceptions of time, but also to their postmodern counterpart: the consecration of uncertainty. Looking at the idealization of the abyss of uncertain futurities and possibility from the perspective of care time, we can wonder if this heroic vision of futurity— expectation or doom—can only become dominant for those whose living infrastructures are taken care of by others.

Care time is also irreducible to productionist time. From the dominant perspective of technoscientific innovation, productivity aims at the economic contribution of practices by the "transformation of materials from a less valued to a *more valued* state" (Thompson 1995, 11). Feminist approaches to care have shown how the work of reproduction and maintenance of life has traditionally been considered marginal to value-creating work, identified to the personal and biological perpetuation (closer to our "animality"). This process can be read from a temporal perspective as all spheres of practice are colonized by the productionist logic; care time is devalued as "unproductive" (Adam 2004, 127) or "merely" reproductive. This seems particularly important for time entangled with the reproduction and maintenance of ecological life. Contributing to thinking the political, economic, and social meanings of care, the ecofeminist philosopher Mary Mellor insightfully articulated an approach to time that emphasizes the importance of "biological time" with regard to production time. Biological time represents for Mellor cycles of the human body, daily needs (sleep, food, excretion, shelter, and clothing) of health and the life cycle. Together with "emotional time," this is time we all need, and though part of it is assumed by social institutions and underpaid work, the substrate of these activities still relies on the "private" world, on what she calls an "immediate altruism" overwhelmingly imposed on women. Drawing upon feminist critiques of this divide, Mellor argued that the world of production as "speeded up" time is only possible because some, albeit dominant, are able to ignore biological and ecological time embeddedness at the expense of women and other carers, as well as the broader ecologies. Productionist time can appear as a separate timescape "in which people do not have to wash their clothes in water full of raw sewage or walk miles to find clean water, fresh fodder or fuelwood. Where people do not have to struggle with heavy shopping bags and small children in pushchairs on and off buses or dash across dangerous roads to get to the school. It is a world that does not have to walk at the pace of the toddling child or the elderly person with emphysema." Idealizing care-time would continue reinforcing this traditionally gendered divide. Mellor notes this as she critiques green utopias based on hope that we "all" would become free of the burdens of productionism, a world based on craft and small-scale technologies. These often fail to think of other work that needs to be done but is relegated or invisible to productionism. For Mellor, this timescape remains the missing link between "high speed lunacy" and the "speed of sustainability" (261). Thinking with care time, I am prolonging the insights of a critique that is now classic, but its transformative potential remains. Within the still predominant productionist timescapes, a politics of care exposes the importance of the work of care for creating livable and lively worlds. Engaging with "care time," I am emphasizing the affectivity and ethicality at stake in life-sustaining doings in all spheres of life—not only those traditionally thought as care work. From the perspective of productionism, time consecrated to the reproduction, maintenance, and repair of ecological soil life, as well as engaging in affective relations with the soil, is wasted time.

Another important aspect of this engagement is resisting the reduction of care work to traditional economic terms (Rose 1994). Valuing care by "efficiency" standards transforms its practice into a managed "conduct" to be monitored (Latimer 2000). That is why, in contexts of managerial control that underestimate care's value and even penalize its practice, acts of care can even be considered as a kind of resistance (Singleton and Law 2013). And yet this is not to say that temporalities of productionism and care do not coexist and capture each other. As I have shown earlier with regard to contemporary soil ecology and alternative models of the living soil, maintaining the very productivity of soil is a strong argument for rejecting intensification and allowing soil renewal. One could even think that the very emphasis in creating relation, the ethical gist of relational ontology, is somehow driven by productionism, as Kathryn Yussof provocatively suggests (Yusoff 2013). Acknowledging the persistence of productionism within alternative notions of soil care while insisting on temporal frictions that disrupt it makes these alternatives communicate from within the hegemonic. Rather than focusing on demonstrating the productive or economic value of activities of care, rather than affirming care as an ideal separate world, and rather than rejecting care as unavoidably implicated, affirming the importance of care time draws attention to, and makes time for, a range of vital practices and experiences that remain discounted, or crushed, or simply unmeasurable in the productionist ethos as we know it and within progressive timescapes of anxious futurity. As a speculative commitment, thinking with care is an obligation contingent to a dominant timescape in which it supports "ethical resistance of the powerless others," by looking out for those, humans and nonhumans, who have the most to lose under productionist-based arrangements.

Finally, perhaps the most disruptive in the ways of making time for soil I have explored is how they transgress the progressive imperative, the "Thou shall not regress" commandment of modern science (Stengers 2012) that still feeds the "innovate or perish" credo. Indeed, the implicit mode of progressive and linear futurity in usual conceptions of innovation could hardly recognize these reconfigurations of soil care. That is why, as Jackson noted, foregrounding the importance of care, maintenance and repair to the very material sustaining of the world is a step in challenging teleological, progressive, shiny ideals of innovation. Care time's irreducibility to productive aims could therefore also contribute to reveal the overestimated value of the productionist imaginary in innovation (Suchman and Bishop 2000).

No output, no growth into the future, no innovation, is possible without a commitment to everyday care. With his focus on the importance of maintenance and repair in technoscience, Steven Jackson calls to disrupt the imaginings of technology that "locate innovation, with its unassailable standing, cultural cachet and valorized economic value, at the top of some change or process, while repair [as a form of care] lies somewhere else: lower, later, or after innovation in process and worth" (Jackson 2014, 227). My approach to care time as both troubling and coexisting with futuristic temporalities is kin to Jackson's diagnosis of the intrinsic connection between dominant notions of innovation and productivism. Jackson calls attention to the unsophisticated worlds of maintenance and repair, which he sees as the work needed to avoid or confront breakdown in information technologies and other endeavors to sustain the world of things. He speaks of a relation to technology that is not only "functional" but "moral" (230), a "very old but routinely forgotten relationship of humans to the things in the world: namely an ethics of mutual care and responsibility."

His plea is for admitting "a possibility denied or forgotten by both the crude functionalism of the technology field and a more traditionally humanist ethics" (231), and therefore he offers a proposition that he recognizes as "tricky":

Is it possible to love, and love deeply, a world of things? Can we bear a substantive ethical, even moral, relationship to categories of objects long consigned to a realm of thin functionalism (a mistake that many of the dominant languages of technology research and design—"usability," "affordances," and so on—tend to reify?). What if we can build new and different forms of solidarity with our objects (and they with us)? And what if, beneath the nose of scholarship, this is what we do every day?" (232)

Jackson's call keeps together care as maintenance and care as affective relation, and brings us back to a politics of caring for neglected "things," which I discussed at the beginning of the book. I bring up these questions here for their relevance to an alterbiopolitics of naturecultural relations in technoscience, to the work of repair and maintenance that care involves in an epoch of ecological hardship—where functionalism and use, as we have seen with the notion of ecosystem services, remain strong. Jackson sees his proposition as "tricky" because of the possibilities of falling into "nostalgia" and "heroism" that he recognizes as challenges for progressive thought (233). These, I might add, are made even trickier by the legacy that progressive critical thought shares with the temporality of futurity—a dread of backwardness that contributes to the almost uniquely accusatory use of "nostalgia." And, importantly, as I will address below, because of the charge of anthropomorphism that hangs over attempts to think the ethical agencies of other than humans.

The charge of backwardness is a heavy inheritance. I mentioned earlier how the underplaying of sophisticated methods give to DIY soil testing an anachronistic aura. The very invocation of involution rather than evolution in the notion of involutionary momentum seems to bring connotations of regression. Speaking of becoming part of the soil community seems to come close to commonly depreciated unscientific spiritual talk (Puig de la Bellacasa 2016). And who has time today for thoughtful and protracted observation anyway? Those who offer such modes of care will be asked to *prove* that they are not nostalgic of an idealized past of immediate connections with nature.[17] And indeed, common reactions to antiproductionist views on technology point at their irrelevance or unviability (read unprofitability) to tackle the important challenges facing current societies: they cannot "feed" the world. So of what kind of science and technology could modes of ecological "unproductive" care time speak of? And what is their relation to futurity and innovation?

Traditional productionist innovation timelines cannot account for these reconfigurations of soil care. Some of these conceptions are deeply untimely because they invoke innovative ways of knowing that will seem inevitably backwards or pre-technoscientific to the progressive spirit. And yet permaculture and agroecology practitioners who engage with foodweb-friendly soil-care techniques describe them as innovations—while simultaneously explaining that some of the "new" technologies that they implement could be a thousand years old, sometimes integrating techniques from contemporary indigenous ecosmologies that claim their ancestrality. This temporal bricolage is not completely absent from contemporary science, as this soil scientist affirms: "The ancient wisdom and indigenous technical knowledge about benefits of manuring, reduced tillage, conservation farming and other practices abandoned somewhere on the way, need to be relearnt" (Rao, in Hartemink 2006, 116). "New" practices recommended by institutions such as the USDA are following similar patterns of reenacting old techniques.[18] These mixed relearnings cannot be understood if we reduce them to a nostalgic return to a preindustrial landscape or one that ignores preindustrial unsustainable soil practices. One can read these interventions as innovative to the current dominant timescape by thinking of them as untimely. They are bringing past doings into a context in which they *become new*; they are innovative *in* the present situation. The reconfiguration of human–soil relations for the inheritors of industrial revolutions is unique to an epoch and timescape where the re-creation of ecological tradition faces global breakdown: a situation that is putting productionism to the test, showing its limits to provide *as well as possible* living conditions, and that requires humans to reconfigure themselves, from soil consumers into soil community members.

Another, less dismissive reading of these temporal redirections would be to see these forms of engagement as refusals of technoscientific mobilization that encourage "slowing down" (Stengers 2005)—in this case, the pace of productionist appropriation of soil life as a resource. Yet the qualification as "slow" could still be misleading. Advocating slowness as time of a different quality against the speed of innovation and growth in technoscience does not necessarily question the progressive direction of the dominant timeline as these approaches do by operating differently within technoscience.[19] The transformative moves in human–soil involvement that I have approached require making time for soil times. Involved soil care poses questions pertaining to relational encounters between coexisting timelines that affect the notions of future in technoscience. In these temporalities of ecological care, growth is not necessarily exponential, nor extensive. This is not only because ecological growth involves cycles of living and dying, but also because what makes a living ecology manifests itself in the intensification and teeming of involvements between members. Conceived as such, the time of soil is not "one"; it exposes multifarious speeds of growth becoming ecologically significant to each other.

Indeed, if we think of time from the perspective of earthworm communities, artificial fertilization of soils aimed at accelerating yield would be a slowing down of the development of worms and other essential soil communities; meanwhile, interventions that adjust to the pace of soil communities' reproductive capacities foster the proliferation and thriving of their habitats. What seems slow or backward when living according to human timeline or timescale might have a different sense in another (Schrader 2010).

And so this speculative journey into care agencies in more than human relations ends up joining calls for decentering unilinear, anthropocentric, temporalities in order to make time for a multiplicity of others. This comes close to the temporal requirements for conceiving nature in the "active voice" (Plumwood 2001). Michelle Bastian shows how the possibilities of attributing significant agency to nonhumans is hindered by a dominant linear conception of time for which change and innovation are only deemed possible for human individual self-directed actions that *break* from the past but remain within a logic of production that requires human control. With Plumwood, Bastian critiques this idea and argues for exposing the unexpected changes, the events, that other than human creative agencies bring to happen. Interestingly, this journey into care time meets diverse temporalities from a different path: not so much that of valuing other than human events and creative ruptures and, therefore as Bastian shows, also disrupting the dominant anthropocentric view of innovation, but by emphasizing uneventful, everyday daily occurrences as transformative. But while the paths might be dissimilar, they concur in affirming what the current dualisms of technoscientific futurity tend to render invisible: the comaking of temporal scales, of the ordinary and the eventful, the so-called reproductive and the productive. In Bastian's words: "It is precisely the repetition necessary to reproduction that opens all living organisms to the ever-present possibility that they might reproduce in ways both unintended and unexpected . . . [the] very possibility that the extraordinary may arise from within the ordinary" (Bastian 2009, 46–47). To argue for a disruption of futuristic time through making care time is therefore not so much about a slowing down of "time," nor a redirection of timelines, but an invitation to rearrange and rebalance relations between a diversity of coexisting temporalities that inhabit the worlds of soil and other interdependent ecologies.

And yet that is why a politics of soil care that insists on perpetuating, maintaining, and intensifying the life of existing cycles involves an ethicopolitical stance on how technoscientific innovation driven by intensification of production and network extension affects relations of care more broadly. As current alarms about the future of soils repeatedly warn, networks that are successful in aligning diverse timelines into the linearity of production endanger the very existence of a living soil and the species that depend on it. Rather than aligning care time to become workable within the dominant timeline—that is, to become productive—the balance of proof is turned toward current ways of living in futurity. How can technoscientific futurity live ecologically with timelines of care? How can sciences and technologies contribute to foster the conditions of material and affective ethicality essential to the living webs of care? These could be relevant questions for disrupting technoscientific futurity. Temporal imaginings that make time for care time contribute to enact a multiplicity of interdependent temporalities, fostering alternatives that challenge the predominance of antiecological timescapes.

Coda

～～～～～～～～～～～～～～～～～～～～～～～～～～～～～～～～～

Across this book I have come back, as a reassuring refrain, to Tronto's generic definition of care. I have also placed it within discussions that engage with more than human worlds and agencies. Tronto stated that care includes "everything that *we* do to maintain, continue, and repair '*our* world'"—our bodies, *our selves,* and our environment—"so that we can live in it as well as possible in a complex, life sustaining web." I have tried to gently decenter the "we" and the "our" that put human agency as the starting point of care, prolonging relational ontologies' ongoing problematization of any claims to a center. I have tried to show ways in which engaging speculatively with a politics of care could further displace the meanings of ethics to respond to the breaking down of modern humanist boundaries. This mode of attention to a more than human life-sustaining web hopefully contributes to efforts in critical posthumanist thought to decenter anthropocentric ethics without discharging humans from specific and situated ethico-political response-abilities (Haraway 2007; Wolfe 2010) required to transform the exploitative relations of anthropocentrism and human exceptionalism. Hopefully too, thinking from the universes of everyday care can help to disrupt the dualistic tale of all humans *versus* all nonhumans that obscures less perceptible ways in which insurgent posthumanist relations are made possible (Papadopoulos 2010). I have also engaged with the ambivalences of caring, connecting interventions that elicit its complexity and often conflictive dimensions. A generic notion of care and the political stance of keeping maintenance work, affectivity, and ethics together was my point of departure for thinking in noncontradictory ways the tensions between decentering human agencies and keeping specific ethical obligations. In turn, the meanings of care become thickened as they are displaced to engage with more than human relations.

And yet, in joining these insightful voices, I still have not fully addressed the expected charge of falling into anthropomorphism in thinking caring as a web of ethicality circulating through agencies involving other than humans. I have attempted to stay close to the more than human web—that is, to relations in which humans are involved—rather than adventuring to speak for other than human only terrains of existence. And yet it is true that I have hinted at the care that comes from other than humans. But can thinking the ethicality of care in this way be more than yet another anthropomorphic delusion—and even another form of anthropocentrism? Or at least, shouldn't I follow my own hesitations about the risks of appropriating others' experiences and refuse to project human-originated ideas of care onto other than humans? Because ethics remains a human thought, can conceptualizing more than human relations of ethicality by disconnecting ethics from individual self-reflective intentionality be enough to address this problem? Is insisting, as I have done, on how other than humans "take care" of the more than human web more than a wishful metaphor? Can there be reciprocity in affective care, however asymmetric, with soil beings? I don't have explanatory answers to such questions. Rather, as I announced in the introduction, if thinking with care in more than human worlds led me to an immersion into a concrete terrain of care practices, ecological human–soil relations, it is precisely this terrain that required the most speculative thought—a thought that I try to reconstruct here as a tentative and provisional conclusion to this journey.

In the introduction, I announced I would be thinking with a tryptic notion of care involving maintenance doings, affective relations, and ethicality as well as political commitment. The disruptive thought of care remains in tension in the uneven interplay of these features. While it might be easier to look past the anthropomorphization of the workers of the soil, the maintenance of biological labors as a way to say that worms "do" some kind of work of care, it is (most probably) true that, *affectively* speaking, worms, nematodes, microbes, and other soil inhabitants do not care about us humans.[1] And indeed, from the exclusively human-centered, morally and politically traditional perspective of the "ethical subject," it would be anthropocentric pretence or, more generously speaking, an interspecies fantasy, to picture a sense of care that circulates proportionally between (all) human and (all) nonhuman beings and material forces. And I am not proposing to settle after this journey on a version of care as some immanent or transcendental mystical force—though I wouldn't see such moves as meaningless for more than human ethics. And yet I am ready to risk the charge of initiating an anthropomorphist ethics of more than human care by blurring the features of care I had myself proposed to delineate. Because for more caring affective ecologies to become possible, we need speculative thinking, and a fair amount of fabulation, so that the anxieties that the attribution of human modes of intentionality to nonhumans generate in critical thinkers do not paralyze our ethical imagination. Barad's questioning of the reduction of ethicality to intentional agency is an obvious inspiration here (Barad 2007). Moreover, the charge of anthropomorphism, as Natasha Myers's work shows by drawing attention to the coforming of knowledge and liveliness that happens when scientists' bodies are affected and transformed by the other than human beings they relate to (Hustak and Myers 2012; Myers 2015), remains within a conception of the other than human world as *passive* and prevents us from opening up to the cues offered by encounters with other than human life forms in which human bodies are also morphed. Care is not one way; the cared for coforms the carer too. Finally, coming back to knowledge politics, if the dislike of anthropomorphism does not prevent scientists, policymakers, and many of us in general from telling stories about the provision of "services" or "functions" by biota such as worms—or naming them "soil managers" or "engineers"—why be shy about disrupting these stories with an imaginary of care? As Haraway puts it, it matters what stories tell stories.

The circulation of care as everyday maintenance of the more than human web of life, conceived as a decentered form of vibrant ethicality, as an ethos rooted in obligations made necessary to specific relations, offers cues to that imagination. A notion of care as a doing rather than a moral intention is the entry point here, but it shouldn't become an impasse. Neglect by contamination circulates through soils, turning a foodweb into a flow of deadly toxicities. Fungi is brought in by activists to feast on polluted soils and clean waste sites in bioremediation efforts. Soil inhabitants and other nonhumans might not be intentionally taking care of human waste to help humans, but the fact is that they do, and it might be said that, within ecological conditions, they can sometimes be immanently obliged to. And in conditions that bring ethical predicaments into the picture, distinctions with coercion can be difficult to draw. Webs of care obligations are impure. Care circulates in all its ambivalence.

The question is how aware soil practitioners, and all of us who benefit from the life in the soil, are to become obliged to worms and other Earth creatures for their work. Thinking these relations through care invites cultivating an ethos grounded in contingent necessities. These obligations are not all equivalent; they are contingent on situated ecological terrains. This journey doesn't add up to a smooth theory of care with no loose ends.

We can be moved to speculative learning with care by multiple stories with which multispecies imaginaries are populating the imaginative desert left by the humanist ethical spirit (Kirksey and Helmreich 2010; Van Dooren 2015; Kirksey 2016). Both from a naturecultural and sociotechnical perspective, we can perceive a web of interdependencies at work that provides conditions for fostering caring ethos and practices. From the sciences to environmental movements, from eco-feminist visions to ethnographies of ecological communities and technoscientific naturecultures, imaginaries of care can help to expose how many other than humans are involved in the agential intra-activities that together make "our" worlds, existences, and doings, and that get earthlings through our interdependent days, taking care of myriad vital processes. Not only thinking these agencies as specific webs of interdependent care helps giving up the notion that these labors are given, mechanical, it also retains the notion that there are specific obligations for those engaged within them as human carers. Across complex life-sustaining webs, the care and the neglect that are put in a world will flow and circulate through living matter and processes. Thinking with care also strengthens the notion that there is no one-fits-all path for the good. What *as well as possible* care might mean will remain a fraught and contested terrain where different arrangements of humans–nonhumans will have different and conflictive significances. In the present of earthly bound naturecultures, the care and neglect that have been put in circulation in the past are still in circulation, effects and consequences transmitted across more than human entanglements.

So while we do not know how to care in advance or once for all, aspiring speculatively for situated ethicalities is vital because no "as well as possible" on Earth is conceivable without these agencies, even those that do not intend themselves as ethical. Situations of care imply nonsymmetrical, multilateral, asubjective, obligations that are distributed across more than human materialities and existences. Thinking with care attracts attention to ethical interrogations meant to seem untimely and worthless from the perspective of predominant unilinear futurities, but we cannot let productivist stories, or even the earnest economies of service, define how nonhuman worlds will be appreciated. There must be other ways to get involved in fostering the ethopoietical liveliness of the more than human agencies that support, currently mostly coercively, that we get the care we need.

May "we" find other ways to be obliged, as well as possible.

Acknowledgments

I am grateful to Florence Degavre, Chloe Deligne, and Nathalie Trussart for encouraging me to think further the meanings of care in knowledge construction and in science as the theme timidly emerged at the end of my doctoral research. Continuing on this path wouldn't have been possible without a European Union Marie Curie Fellowship to undertake postdoctoral research at the University of California, Santa Cruz. Applying to this fellowship only happened thanks to the resolute magic of Sarah Bracke, my dear friend, feminist sister, and coauthor, who helped me believe that we could do it, and the tireless motivation of Didier Demorcy. I am deeply grateful to Isabelle Stengers, who had supervised my bumpy PhD journey with unfaltering and unconditional commitment, for backing the application as well as for her continuous support and adventurous sharing. Donna Haraway not only sponsored my fellowship at the University of California, Santa Cruz, but profoundly affected my thinking with her unique ways of teaching and wholehearted generosity. I will never be able to thank her enough for her mentorship along the way. Not only did she encourage me to develop the ideas in this book, but without having had the chance to be around her extraordinary community, I probably would not have continued on an academic journey.

This book started taking shape during this period. I dearly thank Chris Connery and Gail Hershatter, then directors of the Center for Cultural Studies, University of California, Santa Cruz, for accepting my application as a visiting fellow and welcoming me to this energizing place. To Chris in particular I owe more than I can say in words, not only for our

223

thought-provoking conversations but for being the most fabulous of hosts, taking care of visiting fellows well beyond the call of duty, making Santa Cruz become home. The weekly seminar of the Center was a constant source of exciting ideas that took me out of my comfort zone. In that seminar I first presented the notion of matters of care and received immensely valuable comments and critiques that influenced my understanding of care since then. I thank Jim Clifford, the chair of the Department of History of Consciousness at the time, for cohosting me during these years, giving me access to seminars and classes, and for his stimulating intellectual presence and kindness, always ready to share unreservedly a treasure of wisdom and experience. Mingling with the stunningly smart and welcoming graduate student community at History of Consciousness opened a world for me. I had always thought of myself as doing interdisciplinarity; thanks to them I recogized that I had no idea what that really meant. I learned enormously from conversations in which the personal, political, and intellectual threads of care were inseparable. For this and more, I'd like to thank Harlan Weaver, Jennifer Watanabe, Sha La Bare, and other friends I made during these years at UCSC whose amazing work became crucial for thinking about care. In particular, Natasha Myers, Astrid Schrader, Natalie Loveless, Eben Kirksey, Thom Van Dooren, and Chris Kortright's research and writings remain a continuous source of inspiration, and they have also been generously supportive during this journey. Having the opportunity to collaborate with Jenny Reardon, Jake Metcalf, and others at the UCSC Science and Justice Research Center taught me much about critical friendliness in approaching scientific practices. Karen Barad also took the time to discuss ideas and strongly influenced my thinking. I found a welcoming intellectual home with a science and technology studies community of the San Francisco Bay area and University of California Davis and the exceptionally open and remarkable scholars that make it. Among them, Lochlann Jain, Cori Hayden, Joe Dumit, and Marisol de la Cadena, as well as Rebecca Herzig, also a visiting scholar at the Center for Cultural Studies during my stay, became precious friends to think with. And the very much missed Leigh Star, still so present in so many ways, taught me much about the wonders and troubles of care.

Members of the Groupe d'Études Constructivistes in Brussels, one of the most exciting places to think with that I have ever known, tested in bold conversations the conceptual nuances of Matters of Care. Thinking intimately with Wenda Bauchspies about feminist science studies for a special issue of the journal *Subjectivity* was extremely enriching at many levels and a crucial step in developing the thinking for this book. Joan Haran, Joanna Latimer, Mara Miele, and Rolland Munro, as well as members of the group Culture, Imagination, and Practice at Cardiff University, were extremely supportive and opened my work to other fields of thought. I am also forever grateful to Simon Lilley, head of the School of Management at the University of Leicester, for giving me the opportunity to join a fabulous academic environment in a department where collegial criticality prevails, thanks to his ongoing support. I'd also like to thank Mike Goodman, Ana Viseu, Aryn Martin, Lucy Suchman, Sergio Sismondo, David A. Robinson, and Nicholas Beuret for their ideas and generous comments on earlier versions of chapters. I am extremely grateful to Cary Wolfe for his support and to Joan Tronto for her review and insightful comments on the initial draft of the book.

I dearly thank Clea Kore not only for reading and correcting previous drafts of these chapters but also, with her husband, David Fairchild, for becoming best friends and teachers in joie de vivre. Emma, my dazzling sister, and Marie, my lovely stepmother, thank you for remaining my most indefatigable cheerers. My mother-in-law, Eleni Papadopoulou, who takes care of us and of so many people with solid tenderness, has taught me how starting takes you already half the way—such a precious lesson for a writer. And although not every posthumanist has been a humanist first, this is probably true in my case, thanks to my main caretaker since birth, my constant inspiration, my wonderful, supportive, deeply compassionate father, the humanist scholar Ramón Puig de la Bellacasa. And how to thank my children, Alba and Amaru, the light of my days and of sleepless nights, magical creatures, for keeping me in touch with everyday wonder?

This book is dedicated to my soul companion, without whom I am absolutely certain it would never have materialized, Dimitris Papadopoulos, who read and reread, listened and commented, shouldered and empathized, whose love and care get me through the days with a boundless thirst for joy.

Notes

∻∻∻

Introduction

1. Though now widely used, I first encountered the expression in Marisol de la Cadena's work (2010).

2. I thank Mara Miele for calling my attention to the humanism in this definition.

3. For a distinction between the general and the generic, see Isabelle Stengers (2004).

4. With her engagement with Science Fiction literature, Donna Haraway's work is a particularly salient example of speculative thought as a way to think the future in the present, provoking SF as an interplay between Scientific Fact, Science Fiction, and Speculative Fabulation (Haraway 1992). In this book, the speculative does not play with the epistemology of Scientific Fact as such; the speculative displaces ethics as normative world making (see Haran 2001).

5. One appalling example is the way nurses have been blamed in recent years in the United Kingdom for a "deficit" of care and compassion in the National Health Service leading to worsening conditions for patients across the system. These accusations are framed as a moral issue, while the constant undermining and managerial turn in working conditions for nursing and medical staff remain unaddressed or justified under an efficiency cost-saving principle. Joanna Latimer (2000) offered an illuminating ethnographic study of the effects of managerialism in nursing work. Lara Rachel Cohen offers a radical critique of how the promotion of "compassionate care" in this context diverts attention from critical issues affecting the "body-work" involved in nursing relations of care (Cohen 2011).

6. Katie King speaks of a mode of engaging with our thinking traditions based on a "past-present" temporality, a reenactment that redisposes thoughts and ideas rather than follows a progressive drive to leave them behind.

7. I thank Kobe Matthys for drawing my attention to this quote.

8. "The Radicalisation of Care: Practices, Politics, and Infrastructures," workshop at the Universitat Oberta de Catalunya, November 12–13, 2014. The affirmative tone around the political importance of science and technology studies, and

the aspiration to be involved in transformative politics, is well expressed in the interviews with young researchers gathered in the short film *¿Y si no me lo creo? | What if I don't buy it?*, made by Arianna Mencaroni and Massimiliano Mariotti for the Network for the Social Studies of Science and Technology (esCTS, Spanish STS network), during its third meeting, Barcelona, June 19–21, 2013 (available online on YouTube). I thank Andrea Ghelfi for calling my attention to this film. A recent meeting of this network in July 2015 continued these efforts by exploring "overflows to academic limits," while the 2016 meeting of the Society for Social Studies of Science in Barcelona, "Science and Technology by Other Means," made central a desire of STS interventions beyond academic conversations with the motto "Science and Technology by Other Means."

1. ASSEMBLING NEGLECTED "THINGS"

1. Latour here engages Stengers's thinking of a constructivism that would give up on adjectives (social or philosophical) that result in instituting an explanatory frame (or world) that then pretends to define the terms that hold together the co-construction we affirm as such (that is, when defining a construction as "social" the explanatory frame is allocated to social scientists). See Stengers (2000).

2. Through an effective joke on John Locke's flattening of experience into "the world picture," Latour comments on how the binaries of empiricist epistemology could result from a flat staging aesthetics: a "poor" empiricism that somehow confused ontology with emerging Renaissance *visualization* practices (Latour 2005d, 23).

3. Latour disqualified the use of the explanatory authority of "power" (1993, 125, and 2005b, 85).

4. Though Latour and Harry Collins present very different positions and solutions, they have both shown a common concern with the contribution of the early sociologies of scientific knowledge and social constructivism to a generalized mistrust in facts. Contrary to Latour, Collins is worried by the further undermining of science by the "second wave" of science and technology studies (the ANT focus on nonhuman agency in particular). He argues for more confidence in science, not so much on the epistemological grounds of its superiority, by no means on a political argument of science's strength, but as a "moral" choice.

5. This intervention opens with the scene of the 9/11 attacks in New York City. Latour expressed dismay regarding conspiracy theories that precipitated to debunk the real "causes" of the attack even before the smoke had vanished from the ruins. This preoccupation with the effects of the disrepute of established knowledge is still patent in Latour's latest and maybe most ambitions intervention, the book and collective project *An Inquiry into Modes of Existence* (2013), which starts by staging the challenge with the help of a scene in which a climate scientist confronts climate skeptics not so much by calling them to trust in science as, Latour argues, would have been the argument of a scientist in the past, but to trust institutions again.

6. For the meaning of care as the sharing of a burden, see Tronto (1993, 104–5).

7. Latour has suggested elsewhere the irrelevance of the notion of "standpoint" as essentialist: "standpoints never stand still" (Latour 2000, 380). This critique ignores the complex ways in which the concept of standpoint has been discussed by feminist theorists as a nonessentialist, always moving, contingent, notion. The potential essentialism of standpoints is one of the major unsettled discussions within feminist standpoint thinking. For an anthology of thirty years of debates on the topic, see Harding (2004).

2. Thinking with Care

1. The notion of "diffraction" also responds to debates in STS and the social sciences more largely on how "reflexivity," or reflection, can foster thoughtful and accountable knowledge practices.

2. And this statement can also be read through early Marxist-feminist ontological interventions such as Nancy Hartsock's, in which the world is produced in the interactions of labor (agency) and nature (materiality). Indeed, some of Haraway's early work prolongs socialist-feminist projects (see Haraway 1991 [1978]), and in developing her singular thinking on "naturecultures," Hartsock remains part of this thinking-with web.

3. From *Mille Plateaux* (Deleuze and Guattari 1980, 13), my translation. In Deleuze and Parnet's *Dialogues* (1987, 16), the formula comes back as: "proclaiming 'Long live the multiple' is not yet doing it, one must *do the multiple.*"

4. Paraphrasing Audre Lorde in *Sister Outsider* (1997, 138): "there is no such thing as a single-issue struggle, because we do not live single-issued lives."

5. These conversations do not concern just two authors, yet the dialogue between Haraway and Sandra Harding is particularly significant in this respect. For instance, Haraway's thinking on "situated knowledges" is crafted within a discussion of Harding's framing of the "science question in feminism," while Harding's notion of "strong objectivity" is conceived within a philosophical discussion of "situated knowledges." I have explored the relation between these two thinkers in Puig de la Bellacasa (2004; 2014b).

6. See Harding (2004) for an anthology of these discussions. "Thinking-from" is an illegitimate crossing between a critique of traditional epistemology—as the theory that defines and justifies legitimate grounds for knowledge—and feminist political interventions. As such, to see feminist standpoint theory as merely epistemological theory, a method, or a search for "truth," misses the originality of this connection of theoretical insights and collective practical politics. See, in this sense, the illuminative discussions around Susan Hekman's "Truth or Method," collected in Harding (2004). See also Bracke and Puig de la Bellacasa (2002; 2007).

7. For a beautiful example of how the undecidability of care is at play in practices of auditing and accountability, see Sonja Jerak-Zuiderent's research on performance indicators in health care (2013).

8. Inspired by Samuel Beckett, this phrase was proposed by Stephen Dunne to characterize the mood in UK academia under which the ten-year anniversary

conference of the Centre for Philosophy and Political Economy, School of Management, University of Leicester, was to take place. The statement became the motto of the Conference CPPE@10, Leicester, December 2013.

3. TOUCHING VISIONS

1. The violence of scientific visual technologies has been a classic focus of feminist critique. Ludmilla Jordanova's *Sexual Visions* (1993), for instance, examines the gendered and sexualized aspect of efforts of "seeing" and opening bodies in science and medicine. Important work has focused on visual technologies in antenatal care obstetrics, where the desire of seeing better, the close-up vision of babies in wombs, works for a detachment of the carrying wombs from the mother's body and for the personification of fetuses (Zechmeister 2001).

2. I thank Rebecca Herzig for bringing Thomas Dumm's book to my attention.

3. The image in question is not the main logo of the company anymore, though it can still be found in documents associated with Toltech.

4. "Pope Criticises Pursuit of Wealth!," *BBC News*, October 6, 2008, http://news.bbc.co.uk.

5. FORTIS Bank, Brussels 2008.

6. "Gates Hails Age of Digital Senses," *BBC News*, January 7, 2008, http://news.bbc.co.uk.

7. See Marks (2002), especially chap. 7, "The Logic of Smell," for an account of attempts to commodify the yearnings of our nostrils.

8. *The Economist*, March 8, 2007.

9. For an overview of applications, see Eurohaptics, http://www.eurohaptics.org.

10. Promotional material quoted from http://www.touchpos.co.za/touchpanels.

11. I thank Rebecca Herzig for suggesting this.

12. I am aware of a wealth of nuanced discussions on the relational predicaments of touch in nursing care. A fair account of these discussions would, however, require a level of engagement with the field that goes beyond the scope of this book.

13. A statement of which Arundhati Roy's (2003) version, "Another world is not only possible, she's on the way and, on a quiet day, if you listen very carefully you can hear her breathe," is a beautiful example.

14. For a creative approach to the crafts of virtual handling and grasping as well as to knowing as embodiment, see Natasha Myers's work on the relations of scientists with computer molecule models (Myers 2015).

15. This reminds me of the haptic quality that Deleuze and Guattari (1980, chap. 14) attributed to (nomadic) art, when perception and thinking operate in *smooth* spaces for which there is no preexistent map. While Papadopoulos et al.'s haptic engagement could be read as a prolongation of Deleuze and Guattari, it marks, however, a quite different form of temporality and engagement with everyday experience by breaking with Deleuze and Guattari's fascination with "the event."

16. Politeness as a political art of distance and proximity is beautifully developed in Deleuze's praise of the benevolence inspired by the philosophy and teachings of François Châtelet (Deleuze 2005).

17. Though Gould's argument starts from a very different point of departure (a critique of traditional theories of political obligation in political theory based on free consent between autonomous individuals), she makes a similar point: "Of course the point of the foregoing critique is not to make a simplistic endorsement of women's historical lack of choice as to their 'obligations.' Rather, it is to highlight the necessity for all humans to take care of the activities traditionally assigned to women which do not neatly fit the contract paradigm. Indeed, in institutionalizing the public-private dichotomy and assuming women to such activities—and conversely restricting such activities to 'women's sphere'—these 'obligations' are doubly coercive for women" (Gould 1988, 122–23).

18. I am moving away from reciprocity as a relation between two people to think reciprocity as it may communicate beyond one-to-one relations. A notion of reciprocity as "mutuality" with inspiring consequences for this is also presented by Carol Gould (1983). Gould analyzes various forms of reciprocity (tit for tat or instrumental, reciprocity of respect, etc.) for how they allow a rethinking democratic practice (1988). For a fascinating ethnographic discussion of the tricky dynamics of "taking and giving" as reciprocity in webs of care at play in parental involvement in autism research, see Martine Lappé (forthcoming).

4. Alterbiopolitics

1. Earth Activist Training, http://www.earthactivisttraining.org.

2. Earth Activist Trainings are specific in that they have a strong activist, Direct Action component, focus also on organizing groups in radical democratic ways, and include a spiritual dimension and linked rituals and practices developed in the international neopagan network Reclaiming.

3. I was prompted by Mike Goodman's invitation to participate in his coedited special issue, "Place Geography and the Ethics of Care" (Goodman and McEwan 2010).

4. See, in particular, the Black Permaculture Network, http://blackpermacul turenetwork.org.

5. We could also discuss in this context the success and influence of institutionalized bioethics after World War II (for a posthumanist critique of institutionalized bioethics, see Wolfe 2010).

6. "Life itself" is not simply appropriated; it is made to "collaborate" in its own transformation—and productivity (Cooper 2008).

7. An example of how these moves are entangled in naturecultures is, of course, Haraway's famous cyborg, a hybrid of organic matter and machine materials.

8. Sociopolitical and affective engagements with earthworms exceed the scientific realms of biology. Filippo Bertoni's work is interesting in this regard (Bertoni 2013; Abrahamsson and Bertoni 2014; see also Clark, York, and Bellamy Foster

2009). Conversely, organizational metaphors are also borrowed by biologists and environmental managers to refer to these soil inhabitants, such as "soil engineers" (Lavelle 2000) or even "soil managers" (Sinha et al. 2011). A constant source of inspiration for those fascinated by earthworms remains Charles Darwin's *The Formation of Vegetable Mould, through the Action of Worms, with Observations on Their Habits* (Darwin 1881).

9. I want to dearly thank the Grasshawgs, an international, virtually linked group of doctoral students working with Eben Kirksey, for their generous reading of an earlier version of this chapter and chapter 5 and for their inspiring suggestions. In particular here, Karin Bolender suggested how sometimes beings and things can be better off out of the reach of human care.

10. I thank the reviewer of this manuscript for signaling this inconsistency in the first version.

11. Again, thanks to the Grasshawgs (see note 9 above) for encouraging me to reveal more about the joyful dimension of care and affection within ecological activism in the midst of an atmosphere of fear and doom. Laura McLauchlan, in particular, shared her views on the affective intensities of care that she encounters in her close work with people trying to protect hedgehogs from rapid extinction, confronted with the everyday sadness as they care for these dying creatures. Bringing joy into the thinking of care was also suggested by Suzanna Sawyer, who commented on an earlier version of chapter 5 (at the seminar the Uncommons, organized by Marisol de la Cadena and Mario Blaser at UC Davis, June 2016) and recounted the joy and play of her daughter learning to care for earthworms. What is the joy in care? This is a question that this book has not given enough attention to. I hope others will. It might be because my point of departure has been the reclamation of care by feminist struggles in a context of oppressive relations. But I am really grateful for having been reminded of this because it allowed me to think of my own joyful relation with permaculture practices of care—and others that I am not discussing here, such as caring for children—to at least hint toward the role of joy in becoming affected and caring for.

12. See Latimer and Munro (2009) for a kin narrative of belonging.

5. Soil Times

1. The title of the animation movie by UK-based artist Leo Murray aiming at vulgarizing research on climate change, http://wakeupfreakout.org. For a fascinating study of how the "eco-catastrophic" imaginary reorganized political praxis in environmental movements, see Nicholas Beuret's PhD dissertation, "Organising against the End of the World: The Praxis of Ecological Catastrophe" (2015).

2. The title of the 2015 meeting of the British Sociological Association, "Societies in Transition: Progression or Regression?" sums up our lack of temporal imagination imposed along a logic of linear, unidirectional, progress.

3. Michele Bastian created and coordinates the beautiful research project "Temporal Belongings," which examines the connections between communities and temporality. See http://www.temporalbelongings.org.

4. The return of the dust bowl in contemporary imaginaries is attested by the 2014 blockbuster SF movie *Interstellar*, directed by Christopher Nolan, which depicts the end of the world as a generalized dust bowl; this is accompanied by a worldwide correlative food crisis—though all seen from the U.S. perspective. It also includes accounts inspired by the historical disaster. Interestingly, the lesson here is again a form of technoscientific intensification, but one that is not happening on Earth. We must leave an exhausted Earth to find another planet to terraform. This yearning is well expressed by the restless main protagonist, a space engineer and pilot demoted to the family farm because the world had to leave space exploration adventures to focus on matters of earthly survival. And he hates it: "It's like we've forgotten who we are.... Explorers, pioneers, not *caretakers*.... We're not meant to save the world. We're meant to leave it" (emphasis added).

5. "Land grabbing" refers to the appropriation of land by investors to the detriment of local communities. See http://farmlandgrab.org.

6. In addition to the publications cited in this chapter, see Landa and Feller (2010) and Warkentin (2006). In 1982, the International Union of Soil Sciences set up a working group that led to the establishment of a commission on the History, Philosophy and Sociology of Soil Science.

7. A saying thought to be inspired by Jonathan's Swift's novel *Gulliver's Travels*: "Whoever makes two ears of corn, or two blades of grass to grow where only one grew before, deserves better of mankind, and does more essential service to his country than the whole race of politicians put together."

8. See, for instance, the "Soil Biodiversity Initiative: A Scientific Effort," http://globalsoilbiodiversity.org.

9. Such trends are visible in the information made available to farmers on the USDA Natural Resources Conservation Services on YouTube. For instance, the clip "The Science of Soil Health: Compaction" invites us to "imitate Mother Nature" and limit the use of plowing machinery.

10. See the online petition "Against the 'official' Anthropocene."

11. See http://rodaleinstitute.org.

12. This clip and the following are available on YouTube by searching the cited titles and "Elaine Ingham."

13. "Identifying with Soil Fauna," http://blog.globalsoilbiodiversity.org/article/2013/10/21/identifying-soil-fauna.

14. Schrader (2015) beautifully explores the meanings of care prompted by some students' negative reactions to intimations to "care about" insects deformed by post-Chernobyl radiation—in the face of so many devastated human lives. Schrader discusses "caring about" as an affective relation that does not necessarily involve "caring for" as acting upon a specific need. Exploring the temporalities associated with different ways of experiencing affective care when care is reduced to "direct helping action," Schrader shows how care can be limited by a progressive view oriented to an end or a defined object of care (15). She notes how this type of care is also limited by a logic of exchange and equality that presupposes that this capacity is limited—and therefore accentuates the sense (here in the students'

reactions) that care should be given to humans before "bugs." Instead, Schrader emphasizes a passive dimension of caring about, open-ended care, potentially unlimited, different from active-oriented affectivity.

15. This approach to temporal adjustments resonates with notions of temporal "alignments" explored in STS with relation to collaborative work (Jackson et al. 2011) and analyzed existentially as a process of "torque" by Geoffrey Bowker and Leigh Star (1999). Other processes of technoscientific synchronization in nature-cultures are elicited by Astrid Schrader (2010, 2012).

16. On the eco-ethical importance of multispecies eating together, see Haraway (2008). See also Kristina Lyons (2013) on the specific embodied foodweb conception of soil practitioners in the Colombian Amazonian plains.

17. Paul Kingsnorth has an interesting take on nostalgia that gets at the implications of this common objection. In "Dark Ecology," he argues, "Critics of that book called it nostalgic and conservative, as they do with all books like it. They confused a desire for human-scale autonomy, and for the independent character, quirkiness, mess, and creativity that usually results from it, with a desire to retreat to some imagined 'golden age.' It's a familiar criticism, and a lazy and boring one. Nowadays, when I'm faced with digs like this, I like to quote E. F. Schumacher, who replied to the accusation that he was a 'crank' by saying, 'A crank is a very elegant device. It's small, it's strong, it's lightweight, energy efficient, and it makes revolutions.' Still, if I'm honest, I'll have to concede that the critics may have been on to something in one sense. If you want human-scale living, you doubtless do need to look backward. If there was an age of human autonomy, it seems to me that it probably is behind us. It is certainly not ahead of us, or not for a very long time; not unless we change course, which we show no sign of wanting to do." Available online at https://orionmagazine.org/article/dark-ecology. I thank Nic Beuret for calling my attention to this article.

18. See note 9.

19. See, for instance, *The Slow Science Manifesto*: "Don't get us wrong—we do say yes to the accelerated science of the early 21st century. . . . However, we maintain that this cannot be all. Science needs time to read, and time to fail . . . does not always know what it might be at right now . . . develops unsteadily, with jerky moves and unpredictable leaps forward—at the same time, however, it creeps about on a very slow time scale, for which there must be room and to which justice must be done," http://slow-science.org.

Coda

1. A question for which I have to thank Myra Hird for bringing it forward during discussions at the workshop "The Politics of Care in Technoscience," York University, Toronto, May 2013. Hird's work indeed illuminates human-microbe relations (Hird 2009) as well as the effects of ongoing neglect that manifest in the revolting realities of waste dumping (Hird et al. 2014).

Bibliography

Abbots, Emma-Jayne, Anna Lavis, and Luci Attala, eds. 2015. *Careful Eating: Bodies, Food, and Care.* London: Ashgate.

Abrahamsson, Sebastian, and Filippo Bertoni. 2014. "Compost Politics: Experimenting with Togetherness in Vermicomposting." *Environmental Humanities* 4: 125–48.

Adam, Barbara. 1998. *Timescapes of Modernity: The Environment and Invisible Hazards.* New York: Routledge.

——— 2004. *Time: Key Concepts.* Cambridge: Polity.

Adams, Vincane, Michele Murphy, and Adele E. Clarke. 2009. "Anticipation: Technoscience, Life, Affect, Temporality." *Subjectivity* 28.

Agamben, Giorgio. 1998. *Homo Sacer: Sovereign Power and Bare Life.* Stanford: Stanford University Press.

Ahmed, Nafeez. 2013. "Peak Soil: Industrial Civilisation Is on the Verge of Eating Itself." *Guardian,* June 7.

Ahmed, Sara, and Jackie Stacey. 2001. *Thinking through the Skin (Transformations): Thinking through Feminism.* London: Routledge.

Álvarez Veinguer, Aurora. 2008. "Habitando espacios de frontera: Más allá de la victimización y la idealización de las mujeres migrantes." In *La materialidad de la identidad.* Edited by E. Imaz, 199–219. Donostia: Hariadna.

Atkinson-Graham, Melissa, Martha Kenney, Kelly Ladd, Cameron Michael Murray, and Emily Astra-Jean Simmonds. 2015. "Care in Context: Becoming an STS Researcher." *Social Studies of Science* 45, no. 5: 738–48.

Balfour, Lady Eve. 1943. *The Living Soil.* London: Faber and Faber.

Barad, Karen. 2007. *Meeting the Universe Halfway: Quantum Physics and the Entanglement of Matter and Meaning.* Durham: Duke University Press.

———. 2012. "On Touching—The Inhuman That Therefore I Am." *differences* 23, no. 3: 206–23.

Barbagallo, Camille, and Silvia Federici, eds. 2012. *Care Work and the Commons: The Commoner. A web journal for other values.*

Basel Action Network. 2002. "Exporting Harm: The High-Tech Trashing of Asia," http://www.ban.org.

Bastian, Michelle. 2009. "Inventing Nature: Re-writing Time and Agency in a More-Than-Human World." *Australian Humanities Review: Ecological Humanities Corner* 47 (November): 99–116.

———. 2014. "Time and Community: A Scoping Study." *Time and Society* 23, no. 2: 137–66.

Bauchspies, Wenda, and M. Puig de la Bellacasa, eds. 2009. "Re-tooling Subjectivities: Exploring the Possible with Feminist Science and Technology Studies." *Subjectivity* 28.

Bertoni, Filippo. 2013. "Soil and Worm: On Eating as Relating." *Science as Culture* 22, no. 1: 61–85.

Beuret, Nicholas. 2015. "Organizing against the End of the World: The Praxis of Ecological Catastrophe." PhD diss., School of Management, University of Leicester.

Bial, Raymond. 2000. *A Handful of Dirt*. New York: Walker and Company.

Bird Rose, Deborah. 2012. "Multispecies Knots of Ethical Time." *Environmental Philosophy* 9, no. 1: 127–40.

Bird Rose, Deborah, and Thom Van Dooren, eds. 2011. "Unloved Others: Death of the Disregarded in the Time of Extinctions." *Australian Humanities Review* 50 (May).

Bishop, Hywel. 2010. "The Politics of Care and Transnational Mobility." School of Social Sciences, Cardiff University.

Blackman, Lisa. 2008. *The Body: The Key Concepts*. Oxford: Berg.

Bleier, Ruth, 1984. *Science and Gender: A Critique of Biology and Its Theories on Women*. New York: Pergamon Press.

Boden, Rebecca, Debbie Epstein, and Joanna Latimer. 2009. "Accounting for Ethos or Programmes for Conduct? The Brave New World of Research Ethics Committees." *Sociological Review* 57, no. 4: 727–49.

Boris, Eileen, and Parrenas Rhacel. 2010. *Intimate Labors: Cultures, Technologies, and the Politics of Care*. Stanford: Stanford University Press.

Borras, Saturnino M., Jr., Ruth Hall, Ian Scoones, Ben White, and Wendy Wolford. 2011. "Towards a Better Understanding of Global Land Grabbing: An Editorial Introduction." *Journal of Peasant Studies* 38, no. 2: 209–16.

Borup, Mads, Nik Brown, Kornelia Konrad, and Harro Van Lente. 2006. "The Sociology of Expectations in Science and Technology." *Technology Analysis & Strategic Management* 18, nos. 3–4: 285–98.

Boulaine, Jean. 1994. "Early Soil Science and Trends in Early Literature." In *The Literature of Soil Science*. Edited by Peter McDonald, 20–42. Ithaca: Cornell University Press.

Bouma, J. 2009. "Soils Are Back on the Global Agenda: Now What?" *Geoderma* 150, nos. 1–2: 224–25.

Bouma, J., and Alfred. E. Hartemink. 2002. "Soil Science and Society in the Dutch Context." *Wageningen Journal of Life Sciences* 50, no. 2: 133–40.

Bourg, Julian. 2007. *From Revolution to Ethics*. Montreal: McGill-Queen's University Press.

Bowker, Geoffrey C., and Susan Leigh Star. 1999. *Sorting Things Out: Classification and Its Consequences*. Cambridge, Mass.: MIT Press.

Bracke, Sarah, and Maria Puig de la Bellacasa. 2002. "Building Standpoints." *Tijdschrift vor Genderstudies* 2: 18–29.

———. 2007. "The Arena of Knowledge: Antigone and Feminist Standpoint Theory." In *Doing Gender in Media, Art, and Culture*. Edited by Rosemarie Buikema and Iris van der Tuin, 39–53. New York: Routledge.

Bresnihan, Patrick. 2016. "The More-Than-Human Commons: From Commons to Commoning." In *Space, Power, and the Commons: The Struggle for Alternative Futures*. Edited by S. Kirwan, L. Dawney, and J. Brigstocke. New York: Routledge, 2015.

Brown, Gareth, Emma Dowling, David Harvie, and Keir Milburn. 2012. "Careless Talk: Social Reproduction and Fault Lines of the Crisis in the United Kingdom." *Social Justice: A Journal of Crime, Conflict, and World Order* 39, no. 1: 78–98.

Brown, Nik. 2003. "Hope against Hype: Accountability in Biopasts, Presents, and Futures." *Science Studies* 16, no. 2: 3–21.

Brown, Nik, and Mike Michael. 2003. "A Sociology of Expectations: Retrospecting Prospects and Prospecting Retrospects." *Technology Analysis and Strategic Management* 1, no. 15: 3–18.

Burnett, Graham. 2008. *Permaculture: A Beginner's Guide*. Westcliff on Sea, UK: Spiralseed.

Carlsson, Chriss. 2008. *Nowtopia: How Pirate Programmers, Outlaw Bicyclists, and Vacant-Lot Garderners Are Inventing the Future Today!* Edinburgh: AK Press.

Carrasco, Cristina. 2001. "La sostenibilidad de la vida: Un asunto de mujeres?" *Mientras tanto, Icaria Editorial* 82.

Castañeda, Claudia. 2001. "The Future of Touch." In *Thinking through the Skin*. Edited by Sara Ahmed and Jackie Stacey, 223–36. London: Routledge.

Césaire, Aimé. 2000. *Discourse on Colonialism*. New York: Monthly Review Press.

Charvolin, Florian, André Micoud, and Lynn K. Nyhar, eds. 2007. *Des Sciences citoyennes? La Question de l'amateur dans les sciences naturalists*. La Tour-d'Aigues: Éditions de l'Aube.

Choy, Tim. 2011. *Ecologies of Comparison: An Ethnography of Endangerment in Hong Kong*. Durham: Duke University Press.

Chrétien, Jean-Louis. 2004. *The Call and the Response*. Translated by Anne A. Davenport. New York: Fordham University Press.

Churchman, G. J. 2010. "The Philosophical Status of Soil Science." *Geoderma* 157, nos. 3–4: 214–21.

Clark, B., R. York, and J. Bellamy Foster. 2009. "Darwin's Worms and the Skin of the Earth: An Introduction to Charles Darwin's *The Formation of Vegetable

Mould, through the Action of Worms, with Observations on their Habits (Selections)." *Organization & Environment* 22, no. 3: 338–50.

Clarke, Adele E. 2016. "Anticipation Work: Abduction, Simplification, Hope." In *Boundary Objects and Beyond: Working with Leigh Star*. Edited by Geoffrey C. Bowker, Stefan Timmermans, Adele E. Clarke, and Ellen Balka. Cambridge, Mass.: MIT Press.

Cleaver, Harry M. 1972. "The Contradictions of the Green Revolution." *American Economic Review* 62, nos. 1–2: 177–86.

Cohen, Rachel Lara. 2011. "Time, Space, and Touch at Work: Body Work and Labour Process (Re)Organisation." *Sociology of Health & Illness* 33, no. 2: 189–205.

Coleman, David C., D. A. Crossley, and Paul F. Hendrix. 2004. *Fundamentals of Soil Ecology*. Amsterdam: Elsevier.

Coleman, David C., E. P. Odum, and D. A. Crossley Jr. 1992. "Soil Biology, Soil Ecology, and Global Change." *Biol. Fert. Soils* 14: 104–11.

Collier, Stephen J., and Andrew Lakoff. 2005. "On Regimes of Living." In *Global Assemblages: Technology, Politics, and Ethics as Anthropological Problems*, 22–39. Malden, Mass.: Blackwell.

Collins, H. M., and T. J. Pinch. 1993. *The Golem: What Everyone Should Know about Science*. Cambridge: Cambridge University Press.

Collins, Patricia Hill. 1986. "Learning from the Outsider Within: The Sociological Significance of Black Feminist Thought." *Social Problems* 33, no. 6: S14–S32.

Cooper, Melinda. 2008. *Life as Surplus: Biotechnology and Capitalism in the Neoliberal Era*. Seattle: University of Washington Press.

Cuomo, Chris J. 1997. *Feminism and Ecological Communities: An Ethic of Flourishing*. New York: Routledge.

Curtin, Deane. 1993. "Towards an Ecological Ethic of Care." *Hypatia: A Journal of Feminist Philosophy* 6, no. 1: 60–74.

Danius, Sara. 2002. *The Senses of Modernism: Technology, Perception, and Aesthetics*. Ithaca: Cornell University Press.

Darwin, Charles. 1881. *The Formation of Vegetable Mould, through the Action of Worms, with Observations on their Habits*.

de la Cadena, Marisol. 2010. "Indigenous Cosmopolitics in the Andes: Conceptual Reflections beyond 'Politics.'" *Cultural Anthropology* 25, no. 2: 334–70.

Deleuze, Gilles. 1989. "Qu'est-ce qu'un dispositif?" In *Michel Foucault philosophe*. Edited by François Ewald. Paris: Seuil.

———. 2005. "Pericles and Verdi: The Philosophy of François Châtelet." *Opera Quarterly* 21: 716–24.

Deleuze, Gilles, and Félix Guattari. 1980. *Mille Plateaux*. Paris: Les Éditions de Minuit.

Deleuze, Gilles, and Claire Parnet. 1987. *Dialogues*. New York: Columbia University Press.

———. 1992 [1977]. *Dialogues*. Paris: Flammarion.

Denis, Jerome, and David Pontille. 2015. "Material Ordering and the Care of Things." *Science, Technology & Human Values* 40, no. 3: 338–67.

Dery, Patrick, and Bart Anderson. 2007. "Peak Phosphorus." *Energy Bulletin,* August 13.

Despret, Vincianne. 2004. "The Body We Care For: Figures of Anthropo-zoogenesis." *Body & Society* 10, nos. 2–3: 111–34.

Dewey, John. 1958. *Art as Experience.* New York: Capricorn Books.

Dominguez Rubio, Fernando. 2016. "On the Discrepancy between Objects and Things: An Ecological Approach." *Journal of Material Culture* 21, no. 1: 59–86.

Donovan, Josephine, and Carol J. Adams, eds. 2010. *Beyond Animal Rights: A Feminist Caring Ethic for the Treatment of Animals.* New York: Continuum.

Dowling, Emma. 2012. "The Waitress: On Affect, Method and (Re)presentation." *Cultural Studies–Critical Methodologies* 12, no. 2: 109–17.

Dubinskas, Frank A., ed. 1988. *Making Time: Ethnographies of High-Technology Organizations.* Philadelphia: Temple University Press.

Duffy, Mignon. 2011. *Making Care Count: A Century of Gender, Race, and Paid Care Work.* New Brunswick: Rutgers University Press.

Duffy, Mignon, Amy Armenia, and Clare L. Stacey, eds. 2015. *Caring on the Clock: The Complexities and Contradictions of Paid Care Work.* New Brunswick: Rutgers University Press.

Dumit, Joseph. 2012. *Drugs for Life: How Pharmaceutical Companies Define Our Health.* Durham: Duke University Press.

Dumm, Thomas. 2008. *Loneliness as a Way of Life.* Cambridge, Mass.: Harvard University Press.

Ehrenstein, Amanda. 2006. "Social Relationality and Affective Experience in Precarious Labour Conditions: A Study of Young Immaterial Workers in the Art Industries in Cardiff." School of Social Sciences, Cardiff University.

Engster, Daniel. 2005. "Rethinking Care Theory: The Practice of Caring and the Obligation to Care." *Hypatia: A Journal of Feminist Philosophy* 20, no. 3: 50–74.

———. 2009. *The Heart of Justice: Care Ethics and Political Theory.* Oxford: Oxford University Press.

Esposito, Roberto. 2008. *Bíos: Biopolitics and Philosophy.* Minneapolis: University of Minnesota Press.

Evans, A. B., and M. Miele. 2012. "Between Food and Flesh: How Animals Are Made to Matter (or Not to Matter) within Food Consumption Practices." *Environment and Planning D: Society and Space* 30, no. 2: 298–314.

FAO. 2013. "International Years Council Minutes . . . United Nations: Food and Agricultural Organization of the United Nations."

Fausto Sterling, Anne. 1992. *Myths of Gender: Biological Theories about Women and Men.* New York: Basic Books.

———. 2000. "On Teaching through the Millennium." *Signs: Journal of Women in Culture and Society* 25, no. 4: 1253–56.

Foucault, Michel. 1988. "The Ethic of Care for the Self as a Practice of Freedom. An Interview with Michel Foucault on January 20, 1984." In *The Final Foucault.* Edited by James William Bernauer and David M. Rasmussen, 1–20. Cambridge, Mass.: MIT Press.

———.1990. *The History of Sexuality.* Vol. 3, *The Care of the Self.* London: Penguin.

Frank Peters, Peter. 2006. *Time, Innovation, and Mobilities.* London: Routledge.

Gatzia, Dimitria. 2011. "The Ethics of Care and Economic Theory: A Happy Marriage?" In *Applying Care Ethics to Business.* Issues in Business Ethics 34. Edited by M. Hammington and M. Sander-Staudt. Dordrecht: Springer.

Ghelfi, Andrea. 2015. "The Science Commons: Network Production, Measure, and Organisation in Technoscience." PhD diss., School of Management, University of Leicester.

Gilligan, Carol. 1982. *In a Different Voice: Psychological Theory and Women's Development.* Cambridge, Mass.: Harvard University Press.

Goodman, Michael K. 2013. "iCare capitalism? The Biopolitics of Choice in a Neoliberal Economy of Hope." *International Political Sociology* 7, no. 1: 103–5.

Goodman, Michael K., and Emily Boyd. 2011. "A Social Life for Carbon? Commodification, Markets, and Care." *Geographical Journal* 177, no. 2: 102–9.

Goodman, Mike, and Cheryl McEwan, eds. 2010. "Place, Geography, and the Ethics of Care." Special issue of *Ethics, Place & Environment: A Journal of Philosophy & Geography* 13, no. 2.

Gould, Carol. 1983. "Beyond Causality in the Social Sciences: Reciprocity as a Model of Non-Exploitative Social Relations." In *Epistemology, Methodology, and the Social Sciences: Boston Studies in the Philosophy of Science.* Edited by R. S. Cohen and M. W. Wartofsky. Dordrecht: Springer.

———.1988. *Rethinking Democracy.* Berkeley: University of California Press.

Guattari, Félix. 2000. *The Three Ecologies.* London: Athlone.

Hankivsky, Olena. 2004. *Social Policy and the Ethic of Care.* Vancouver: University of British Columbia Press.

Haran, Joan. 2001. "Why Turn to Speculative Fiction? On Reconceiving Feminist Research for the Twenty-First Century." In *Gender, Health, and Healing: The Public/Private Divide.* Edited by Gillian Bendelow et al., 68–87. London: Routledge.

———. 2010. "Redefining Hope as Praxis." *Journal for Cultural Research* 14, no. 3: 393–408.

Haraway, Donna. 1991a. "A Cyborg Manifesto: Science, Technology, and Socialist-Feminism in the Late Twentieth Century." In *Simians, Cyborgs, and Women: The Reinvention of Nature,* 149–81. New York: Routledge.

———.1991b. "In the Beginning Was the Word: The Genesis of Biological Theory." In *Simians, Cyborgs, and Women.* New York: Routledge.

———.1991c. *Simians, Cyborgs, and Women.* New York: Routledge.

———. 1991d. "Situated Knowledges: The Science Question in Feminism and the Privilege of Partial Perspective." In *Simians, Cyborgs, and Women,* 183–201. New York: Routledge.

———. 1991 [1978]. "Animal Sociology and a Natural Economy of the Body Politic: A Political Physiology of Dominance." In *Simians, Cyborgs, and Women.* New York: Routledge.

———. 1992. *Primate Visions: Gender, Race, and Nature in the World of Modern Science.* London: Verso.

———. 1994a. "A Game of Cat's Cradle: Science Studies, Feminist Theory, Cultural Studies." *Configurations* 1: 59–72.

———. 1994b. "Never Modern, Never Been, Never Ever: Some Thoughts about Never-Never Land in Science Studies." Paper presented at the 4S Meeting, New Orleans.

———. 1997a. *Modest_Witness@Second_Millennium. FemaleMan©_Meets_Onco-Mouse™: Feminism and Technoscience.* New York: Routledge.

———. 1997b. *Modest_Witness@Second_Millennium.FemaleMan_Meets_OncoMouse: Feminism and Technoscienc.* New York: Routledge.

———. 2000. "Morphing in the Order: Flexible Strategies, Feminist Science Studies, and Primate Revisions." In *Primate Encounters: Models of Science, Gender, and Society.* Edited by Shirley C. Strum and Linda Marie Fedigan. Chicago: University of Chicago Press.

———. 2003. *The Companion Species Manifesto: Dogs, People, and Significant Otherness.* Chicago: Prickly Paradigm Press.

———. 2007. *When Species Meet.* Minneapolis: University of Minnesota Press.

———. 2011. "Speculative Fabulations for Technoculture's Generations: Taking Care of Unexpected Country." In *Australian Humanities Review* 50. Edited by Deborah Bird Rose and Thom Van Dooren.

———. 2015. "Anthropocene, Capitalocene, Plantationocene, Chthulucene: Making Kin." *Environmental Humanities* 6: 159–65.

———. 2016. *Staying with the Trouble: Making Kin in the Chthulucene.* Durham: Duke University Press.

Haraway, Donna, and Thyrza Nichols Goodeve. 2000. *How Like a Leaf: Donna J. Haraway; An Interview with Thyrza Nichols Goodeve.* New York: Routledge.

Harding, Sandra. 1986. *The Science Question in Feminism.* Ithaca: Cornell University Press.

———. 1991. *Whose Science? Whose Knowledge? Thinking from Women's Lives.* Ithaca: Cornell University Press.

———, ed. 2004. *The Feminist Standpoint Theory Reader: Intellectual and Political Controversies.* New York: Routledge.

———. 2008. *Sciences from Below: Feminisms, Postcolonialities, and Modernities.* Durham: Duke University Press.

Hardt, Michael, and Antonio Negri. 2000. *Empire.* Cambridge, Mass.: Harvard University Press.

———. 2009. *Commonwealth.* Cambridge, Mass.: Belknap Press of Harvard University Press.

Hartemink, Alfred E., ed. 2006. *The Future of Soil Science, CIP–Gegevens Koninklijke Bibliotheek, Den Haag.* Wageningen: IUSS Union of Soil Sciences.

———. 2008. "Soils Are Back on the Global Agenda." *Soil Use and Management* 24, no. 4: 327–30.

Hartemink, Alfred E., and Alex McBratney. 2008. "A Soil Science Renaissance." *Geoderma* 148, no. 2: 123–29.

Hartsock, Nancy. 1983. "The Feminist Standpoint: Toward a Specifically Feminist Historical Materialism." In *Money, Sex, and Power: Toward a Feminist Historical Materialism.* Edited by Nancy Hartsock, 231–51. New York: Longman.

Hayward, Eva. 2010. "'Fingeryeyes': Impressions of Cup Corals." *Cultural Anthropology* 4, no. 24: 577–99.

Hedgecoe, Adam, and Paul Martin. 2003. "The Drugs Don't Work: Expectations and the Shaping of Pharmacogenetics." *Social Studies of Science* 33, no. 3: 327–64.

Helms, Douglas. 1997. "Land Capability Classification: The U.S. Experience." In *History of Soil Science: International Perspectives.* Edited by D. H. Yaalon, 159–75. Reiskirchen: Catena Verlag.

Hess, David J. 1997. *Science Studies: An Advanced Introduction.* New York: New York University Press.

———. 2007. "Crosscurrents: Social Movements and the Anthropology of Science and Technology." *American Anthropologist* 109, no. 3: 463–72.

Heyes, Cressida J. 2007. *Self-Transformations: Foucault, Ethics, and Normalized Bodies.* New York: Oxford University Press.

Hillel, Daniel. 1992. *Out of the Earth: Civilization and the Life of the Soil.* Berkeley: University of California Press.

———, ed. 2004. *Encyclopedia of Soils in the Environment,* vol. 1. Amsterdam: Elsevier/Academic Press.

Hird, M. J., S. Lougheed, R. K. Rowe, and C. Kuyvenhoven. 2014. "Making Waste Management Public (or Falling Back to Sleep)." *Social Studies of Science* 44, no. 3: 441–65.

Hird, Myra J. 2009. *The Origins of Sociable Life: Evolution after Science Studies.* Houndmills, Basingstoke: Palgrave Press.

Hoagland, Sarah L. 1991. "Some Thoughts about Caring." In *Feminist Ethics: New Essays.* Edited by Claudia Card, 246–63. Lawrence: University of Kansas Press.

Hochschild, Arlie Russell. 1983. *The Managed Heart: Commercialization of Human Feeling.* Berkeley: University of California Press.

Hole, Francis D. 1988. "The Pleasures of Soil Watching." *Orion Nature Quarterly* (Spring): 6–11.

Holmgren, David. 2002. *Permaculture: Principles and Pathways Beyond Sustainability.* Hepburn, Victoria: Holmgren Design Services.

hooks, bell. 2000. *All about Love: New Visions.* New York: William Morrow.

Hoy, David. 2004. *Critical Resistance: From Poststructuralism to Post-Critique.* Cambridge, Mass.: MIT Press.

Hustak, Carla, and Natasha Myers. 2012. "Involutionary Momentum: Affective Ecologies and the Sciences of Plant/Insect Encounters." *differences: A Journal of Feminist Cultural Studies* 23, no. 3: 74–117.

Ingham, Elaine. 2002. *The Compost Tea Brewing Manual,* 5th ed. Bentley: Soil Foodweb Institute.

———.2004. "The Soil Foodweb: Its Role in Ecosystems Health." In *The Overstory Book: Cultivating Connections with Trees.* Edited by Craig R. Elevitch, 62–65. Holualoa, Hawaii: Permanent Agriculture Resources.

Jackson, Steven J. 2014. "Rethinking Repair." In *Media Technologies: Essays on Communication, Materiality, and Society.* Edited by Tarleton Gillespie, Pablo Boczkowski, and Kirsten Foot. Cambridge Mass.: MIT Press.

Jackson, Steven J., and Laewoo Kang. 2014. "Breakdown, Obsolescence, and Reuse: HCI and the Art of Repair," 449–58. *Proceedings of the SIGCHI Conference on Human Factors in Computing Systems.* New York: ACM.

Jackson, Steven J., David Ribes, Ayse G. Buyuktur, and Geoffrey C. Bowker. 2011. "Collaborative Rhythm: Temporal Dissonance and Alignment in Collaborative Scientific Work." Computer Supported Cooperative Work, Hangzhou, China, March 19–23.

Jackson, Sue, and Lisa R. Palmer. 2015. "Reconceptualizing Ecosystem Services: Possibilities for Cultivating and Valuing the Ethics and Practices of Care." *Progress in Human Geography* 39, no. 2: 122–45.

Jaggar, Alison M. 2001. "Feminist Ethics." In *The Blackwell Guide to Ethical Theory.* Edited by H. LaFollette, 348–74. Oxford: Blackwell.

Jain, S. Lochlann. 2006. *Injury: The Politics of Product Design and Safety Law in the United States.* Princeton: Princeton University Press.

———.2013. *Malignant: How Cancer Becomes Us.* Berkeley: University of California Press.

Jay Gould, Stephen. 1987. *Time's Arrow, Time's Cycle: Myth and Metaphor in the Discovery of Geological Time.* Cambridge, Mass.: Harvard University Press.

Jensen, Derrick. 2009. "Forget Shorter Showers." *Orion Magazine,* July, https://orionmagazine.org/article/forget-shorter-showers/.

Jerak-Zuiderent, Sonja. 2013. "Generative Accountability: Comparing with Care." Institute of Health Policy and Management, Erasmus University, Rotterdam.

———.2015. "Accountability from Somewhere and for Someone: Relating with Care." *Science as Culture* 24, no. 4: 412–35.

Johnson, Deborah G., and James M. Wetmore. 2008. "STS and Ethics: Implications for Engineering Ethics." In *The Handbook of Science and Technology Studies,* 3rd ed. Edited by Edward J. Hackett, Olga Amsterdamska, Michael Lynch, and Judy Wajcman. Cambridge, Mass.: MIT Press.

Johnson, Leigh. 2010. "The Fearful Symmetry of Arctic Climate Change: Accumulation by Degradation." *Environment and Planning D: Society and Space,* no. 5: 828–47.

Jones, Campbell, Martin Parker, and Rene Ten Bos. 2005. *For Business Ethics.* London: Routledge.

Jordanova, Ludmilla. 1993. *Sexual Visions: Images of Gender in Science and Medicine between the Eighteenth and Twentieth Centuries.* Madison: University of Wisconsin Press.

Keller, Evelyn F. 1984. *A Feeling for the Organism: Life and Work of Barbara McClintock*. New York: W. H. Freeman.

———. 1985. *Reflections on Gender and Science*. New Haven: Yale University Press.

Kershaw, Paul. 2005. *Carefair: Rethinking the Responsibilities and Rights of Citizenship*. Vancouver: University of British Columbia Press.

Keulartz, Joseph, Maartje Schermer, Michiel Korthals, and Tsalling Swierstra. 2004. "Ethics in Technological Culture: A Programmatic Proposal for a Pragmatist Approach." *Science, Technology & Human Values* 29, no. 1: 3–29.

King, Katie. 2012. *Networked Reenactments: Stories Transdisciplinary Knowledges Tell*. Durham: Duke University Press.

King, Roger J. H. 1991. "Caring about Nature: Feminist Ethics and the Environment." *Hypatia: A Journal of Feminist Philosophy* 6, no. 1: 75–89.

Kirksey, S. Eben. 2015. *Emergent Ecologies*. Durham: Duke University Press.

Kirksey, S. Eben, and Stefan Helmreich. 2010. "The Emergence of Multispecies Ethnography." *Cultural Anthropology* 25, no. 4: 545–76.

Kittay, Eva Feder. 1999. *Love's Labor: Essays on Women, Equality, and Dependency*. New York: Routledge.

Kittay, Eva Feder, and Ellen K. Feder. 2002. *The Subject of Care: Feminist Perspectives on Dependency*. Lanham, Md.: Rowman & Littlefield.

Kortright, Chris. 2013. "On Labor and Creative Transformations in the Experimental Fields of the Philippines." *East Asian Science, Technology and Society: An International Journal*, no. 7: 557–78.

———. 2015. "From Doomsday to Promise: Visions of Evolution in C4 Rice." In *International Rice Research and Development: 50 years of IRRI for Global Food Security, Stability, and Welfare*. Edited by Margreet van der Burg and Harro Maat. New York: CABI.

Kröger, Teppo. 2009. "Care Research and Disability Studies: Nothing in Common?" *Critical Social Policy* 29, no. 3: 398–420.

Krupenikov, Igor Arcadie. 1993. *History of Soil Science: From Its Inception to Present*. New Delhi: Amerind Publishing Co.

Landa, Edward R., and Christian Feller. 2010. *Soil and Culture*. Dordrecht: Springer Science+Business Media.

Lappé, Martine. 2014. "Taking Care: Anticipation, Extraction, and the Politics of Temporality in Autism Science." *Biosocieties* 9, no. 3: 304–28.

Latimer, Joanna. 2000. *The Conduct of Care: Understanding Nursing Practice*. London: Blackwell.

Latimer, Joanna, and Mara Miele, eds. 2013. "Naturecultures? Science, Affect, and the Non-Human." Special Issue of *Theory, Culture and Society* 30: 7–8.

Latimer, Joanna, and Rolland Munro. 2009. "Keeping and Dwelling: Relational Extension, the Idea of Home, and Otherness." *Space and Culture* 12, no. 3: 317–31.

Latimer, Joanna, and Maria Puig de la Bellacasa. 2013. "Re-thinking the Ethical in Bioscience: Everyday Shifts of Care in Biogerontology." In *Re-theorising the Ethical*. Edited by Nicky Priaulx. London: Ashgate.

Latour, Bruno. 1987. *Science in Action: How to Follow Scientists and Engineers through Society.* Cambridge, Mass.: Harvard University Press.

———. 1993. *We Have Never Been Modern.* Cambridge, Mass.: Harvard University Press.

———. 1996a. *Aramis, or the Love of Technology.* Cambridge, Mass.: Harvard University Press.

———. 1996b. *Petite réflexion sur le culte moderne des dieux faitiches* [sic], Collection *Les empêcheurs de penser en rond.* Le Plessis-Robinson: Synthélabo groupe.

———. 1999. *Pandora's Hope: Essays on the Reality of Science Studies.* Cambridge, Mass.: Harvard University Press.

———. 2000. "A Well-Articulated Primatology: Reflections of a Fellow Traveller." In *Primate Encounters: Models of Science, Gender, and Society.* Edited by Shirley C. Strum and Linda Marie Fedigan. Chicago: University of Chicago Press.

———. 2004a. *Politics of Nature: How to Bring the Sciences into Democracy.* Cambridge: Mass.: Harvard University Press.

———. 2004b. "Why Has Critique Run Out of Steam? From Matters of Fact to Matters of Concern." *Critical Inquiry* 30, no. 2: 225–48.

———. 2005a. "From Realpolitik to Dingpolitik, or How to Make Things Public." In *Making Things Public: Atmospheres of Democracy.* Edited by Bruno Latour and Peter Weibel, 14–43. Cambridge, Mass.: MIT Press.

———. 2005b. *Reassembling the Social: An Introduction to Actor-Network Theory.* New York: Oxford University Press.

———. 2005c. "Victor Frankenstein's Real Sin." *Domus* (February): 878.

———. 2005d. "What Is the Style of Matters of Concern? Two Lectures in Empirical Philosophy." Spinoza Chair in Philosophy Lectures at the University of Amsterdam, April–May 2005.

———. 2007a. "'It's development, stupid!' or, How to Modernize Modernization," http://www.bruno-latour.fr.

———. 2007b. "Turning Around Politics: A Note on Gerard de Vries's Paper." *Social Studies of Science* 37, no. 5: 811–20.

———. 2013. *An Inquiry into Modes of Existence: An Anthropology of the Moderns.* Translated by Catherine Porter. Cambridge, Mass.: Harvard University Press.

Lavelle, Patrick. 2000. "Ecological Challenges for Soil Science." *Soil Science* 165, no. 1: 73–86.

Lavelle, Patrick, and Alister V. Spain. 2003. *Soil Ecology.* Dordrecht: Kluwer Academic.

Lezaun, Javier, and Steve Woolgar, eds. 2013. "A Turn to Ontology in Science and Technology Studies?" Special issue of *Social Studies of Science* 43, no. 3.

Lilley, S., and D. Papadopoulos. 2014. "Material Returns: Cultures of Valuation, Biofinancialisation, and the Autonomy of Politics." *Sociology* 48, no. 5: 972–88.

Lohan, Tara. 2008. *Water Consciousness: How We All Have to Change to Protect Our Most Critical Resource.* San Francisco: Healdsburg.

López Gil, Silvia. 2007. "Las lógicas del cuidado." *Diagonal* 50.

Lorde, Audre. 1997. *Sister Outsider: Essays and Speeches*. Freedom, Calif.: Crossing Press.

Lykke, Nina. 1996. "Between Monsters, Goddesses, and Cyborgs." In *Between Monsters, Goddesses, and Cyborgs: Feminist Confrontations with Science, Medicine, and Cyberspace*. Edited by Nina Lykke and Rosi Braidotti. London: ZED Books.

Lyons, Kristina Marie. 2013. "Soil Practitioners and 'Vital Spaces': Agricultural Ethics and Life Processes in the Colombian Amazon." Department of Anthropology, University of California, Davis.

———. 2014. "Soil Science, Development, and the 'Elusive Nature' of Colombia's Amazonian Plains." *Journal of Latin American and Caribbean Anthropology* 19, no. 2: 212–36.

———. 2016. "Decomposition as Life Politics: Soils, *Selva*, and Small Farmers under the Gun of the U.S.-Colombia War on Drugs." *Cultural Anthropology* 31, no. 1: 56–81.

Malos, E., ed. 1980. *The Politics of Housework*. London: Allison and Busby.

Marion Young, Iris. 1997. "Asymmetrical Reciprocity: On Moral Respect, Wonder, and Enlarged Thought." *Constellations* 3, no. 3.

———. 2000. *Inclusion and Democracy*. Oxford: Oxford University Press.

Marks, Laura U. 2002. *Touch: Sensuous Theory and Multisensory Media*. Minneapolis: University of Minnesota Press.

Marres, Noortje. 2007. "The Issues Deserve More Credit: Pragmatist Contributions to the Study of Public Involvement in Controversy." *Social Studies of Science* 37, no. 5: 59–80.

Martin, Aryn, Natasha Myers, and Ana Viseu, eds. 2015. "The Politics of Care in Technoscience." Special Issue of *Social Studies of Science*.

Martin, Brian. 1993. "The Critique of Science Becomes Academic." *Science, Technology & Human Values* 18, no. 2: 247–59.

Mayberry, Maralee, Banu Subramaniam, and Lisa Weasel, eds. 2001. *Feminist Science Studies: A New Generation*. New York: Routledge.

Mbembé, Achille J. 2001. *On the Postcolony*. Berkeley: University of California Press.

McCaffrey, Anne. 1991. *The Ship Who Sang*. Del Rey: Ballantine.

McDonagh, John. 2014. "Rural Geography II: Discourses of Food and Sustainable Rural Futures." *Progress in Human Geography* 38, no. 6: 838–44.

McDonald, P. 1994. "Characteristics of Soil Science Literature." In *The Literature of Soil Science*. Edited by P. McDonald, 43–72. Ithaca: Cornell University Press.

Mellor, Mary. 1997. *Feminism and Ecology*. Cambridge: Polity Press.

Mendum, Ruth. 2009. "Subjectivity and Plant Domestication: Decoding the Agency of Vegetable Food Crops." *Subjectivity* 28: 316–33.

Metcalf, Jake, and Thom Van Dooren, eds. 2012. "Temporal Environments: Rethinking Time and Ecology." Preface. Special Issue of *Environmental Philosophy* 9, no. 1: v–xiv.

Metzger, Jonathan. 2013. "Placing the Stakes: The Enactment of Territorial Stake-holders in Planning Processes." *Environment and Planning A* 45: 781–96.

———. 2014. "Spatial Planning and/as Caring for More-Than-Human Place." *Environment and Planning A* 46, no. 5: 1001–11.

Michael, Mike. 2001. "Futures of the Present: From Performativity to Prehension." In *Contested Futures: A Sociology of Prospective Techno-Science*. Edited by Nik Brown, Brian Rappert, and Andrew Webster. Aldershot: Ashgate.

Millennium Ecosystem Assessment. 2005. *Ecosystems and Human Well-Being: Synthesis*. Washington, D.C.: Island Press.

Mol, Anne Marie. 1999. "Ontological Politics: A World and Some Questions." In *Actor Network Theory and After*. Edited by John Law and John Hassard. Oxford: Blackwell.

———. 2008. *The Logic of Care: Health and the Problem of Patient Choice*. New York: Routledge.

Mol, Anne Marie, Ingunn Moser, and Jeannette Pols, eds. 2010. *Care in Practice: On Tinkering in Clinics, Homes, and Farms*. Bielefeld: Transcript.

Mollison, Bill. 1988. *Permaculture: A Designer's Manual*. Sisters Creek/Tasmania: Tagari Publications.

Monbiot, George. 2015. "We're Treating Soil Like Dirt. It's a Fatal Mistake, as Our Lives Depend on It." *Guardian*, March 25.

Moore, Jason. 2010. "The End of the Road: Agricultural Revolutions in the Capitalist World-Ecology, 1450–2010." *Journal of Agrarian Change* 10, no. 3: 389–413.

———. 2014. "The Capitalocene I: On the Nature and Origins of Our Ecological Crisis." *Jason W. Moore. Essays, Articles and Interviews:* http://www.jasonwmoore.com/Essays.html.

Morton, Timothy. 2009. *Ecology without Nature: Rethinking Environmental Aesthetics*. Cambridge, Mass.: Harvard University Press.

Muller, Ruth. 2012. "On Becoming a 'Distinguished' Scientist: Careers, Individuality, and Collectivity in Postdoctoral Researchers' Accounts on Living and Working in the Life Sciences." Sociology/Science and Technology Studies, University of Vienna.

Muller, Ruth, and Martha Kenney. 2014. "Agential Conversations: Interviewing Postdoctoral Life Scientists and the Politics of Mundane Research Practices." *Science as Culture* 23, no. 4: 537–59.

Munro, Rolland. 2005. "Partial Organization: Marilyn Strathern and the Elicitation of Relations." In *Contemporary Organization Theory*. Edited by Campbell Jones and Rolland Munro. Oxford: Blackwell.

Münster, Daniel. 2015. "'Ginger is a gamble': Crop Booms, Rural Uncertainty, and the Neoliberalization of Agriculture in South India." *Focaal* 71: 100–113.

———. Forthcoming. "Cultivating Hope in South India: The Affective Ecologies of Natural Farming." In *Affective Ecologies*. Edited by Neera Singh. Tucson: University of Arizona Press.

Murphy, M. 2015. "Unsettling Care: Troubling Transnational Itineraries of Care in Feminist Health Practices." *Social Studies of Science* 45, no. 5: 717–37.

Myers, Natasha. 2008. "Molecular Embodiments and the Body-Work of Modelling in Protein Crystallography." *Social Studies of Science* 32, no. 2: 163–99.

———. 2013. "Sensing Botanical Sensoria: A Kriya for Cultivating Your Inner Plant." Centre for Imaginative Ethnography, http://imaginativeethnography.org/imag inings/affect/sensing-botanical-sensoria.

———. 2015. *Rendering Life Molecular: Modeling Proteins and Making Scientists in the Twenty-First Century Life Sciences*. Durham: Duke University Press.

Myers, Natasha, and Joseph Dumit. 2011. "Haptic Creativity and the Mid-embodiments of Experimental Life." In *A Companion to the Anthropology of the Body and Embodiment*. Edited by Frances E. Mascia-Lees. Malden, Mass.: Wiley-Blackwell.

Nair, Indira. 2001. "Science and Technology with Care: Structuring Science in the Framework of Care, Multiplicity, and Integrity." *Journal of College Science Teaching* 30, no. 4: 274–77.

Nixon, Rob. 2011. *Slow Violence and the Environmentalism of the Poor*. Cambridge Mass.: Harvard University Press.

Noddings, Nel. 1984. *Caring: A Feminine Approach to Ethics and Moral Education*. Berkeley: University of California Press.

Ong, Aihwa, and Stephen J. Collier. 2005. *Global Assemblages: Technology, Politics, and Ethics as Anthropological Problems*. Malden, Mass.: Blackwell.

Oudshoorn, Nelly. 2008a. "Diagnosis at a Distance: The Invisible Work of Patients and Healthcare Professionals in Cardiac Telemonitoring Technology." *Sociology of Health and Illness* 30, no. 2: 272–88.

———. 2008b. "Acting with Telemonitoring Technologies." Paper delivered at the Annual Meeting of the Society for Social Studies of Science, Rotterdam, August 22.

Palli Monguillod, C. 2004. "Entangled Laboratories: Liminal Practices in Science." PhD diss., Universitat Autonoma de Barcelona, at http://dialnet.unirioja.es/servlet/tesis?codigo=5680.

Papadopoulos, Dimitris. 2010. "Insurgent Posthumanism." *Ephemera: Theory & Politics in Organization* 10, no. 2: 134–51.

———. 2011. "Alter-ontologies: Towards a Constituent Politics in Technoscience." *Social Studies of Science* 41, no. 2: 177–201.

———. 2014a. "From Publics to Practitioners: Invention Power and Open Technoscience." *Science as Culture* 24, no. 1: 108–21.

———. 2014b. "Politics of Matter: Justice and Organisation in Technoscience." *Social Epistemology* 28, no. 1: 70–85.

———. 2017. *Experimental Politics: Technoscience and More Than Social Movements*. Durham: Duke University Press.

Papadopoulos, Dimitris, Niamh Stephenson, and Vassilis Tsianos. 2008a. *Escape Routes: Control and Subversion in the Twenty-First Century*. London: Pluto Press.

Paterson, Mark. 2006. "Feel the Presence: Technologies of Touch and Distance." *Environment and Planning D: Society and Space* 24: 691–708.

———. 2007. *The Senses of Touch: Haptics, Affects, and Technologies.* Oxford: Berg.

Penley, Constance, Andrew Ross, and Donna Haraway. 1990. "Cyborgs at Large. An Interview with Donna Haraway." *Social Text* 25/26: 8–22.

Perez-Bustos, Tania. 2014. "Of Caring Practices in the Public Communication of Science: Seeing through Trans Women Scientists' Experiences." *Signs: Journal of Women in Culture and Society* 39, no. 4: 857–66.

Petryna, Adriana. 2009. *When Experiments Travel: Clinical Trials and the Global Search for Human Subjects.* Princeton: Princeton University Press.

Pignarre, Philippe, and Isabelle Stengers. 2011. *Capitalist Sorcery: Breaking the Spell.* London: Palgrave.

Pimm, Stuart L., John H. Lawton, and Joel E. Cohen. 1991. "Food Web Patterns and Their Consequences." *Nature* 350: 669–74.

Plumwood, Val. 2001. "Nature as Agency and the Prospects for a Progressive Naturalism." *Capitalism Nature Socialism* 12, no. 4: 3–32.

Povinelli, Elizabeth A. 2011. "The Governance of the Prior." *Interventions* 13, no. 1: 13–30.

Precarias a la Deriva. 2004. *A la deriva, por los circuitos de la precariedad femenina.* Madrid: Traficantes de Sueños.

———. 2006. "A Very Careful Strike." *The Commoner* 11.

Puig de la Bellacasa, Maria. 2004. "Think We Must: Feminist Politics and the Construction of Knowledge." PhD diss., Department of Philosophy, Université Libre de Bruxelles.

———. 2013. *Politiques féministes et construction des savoirs. "Penser nous devons"!* Paris: L'Harmattan.

———. 2014a. "Encountering Bioinfrastructure: Ecological Struggles and the Sciences of Soil." *Social Epistemology* 28, no. 1: 26–40.

———. 2014b. *Les savoirs situés de Sandra Harding et Donna Haraway: Science et épistémologies féministes.* Paris: L'Harmattan.

———. 2016. "Ecological Thinking and Material Spirituality." In *Boundary Objects and Beyond: Working with Leigh Star.* Edited by Geoffrey C. Bowker, Stefan Timmermans, Adele E. Clark, and Ellen Balka, 47–68. Cambridge, Mass.: MIT Press.

Radcliffe, Matthew. 2008. "Touch and Situatedness." *International Journal of Philosophical Studies* 16, no. 3: 299–322.

Raffles, Hugh. 2010. *Insectopedia.* New York: Pantheon Books.

Renz, Katie. 2003. "Cultivating Hope at Earth Activist Training." *Hopedance* 39.

Richter, Daniel deB, Allan R. Bacon, and L. Mobley Megan et al. "Human–Soil Relations Are Changing Rapidly: Proposals from SSSA's Cross-Divisional Soil Change Working Group." *Soil Science Society of America Journal* 75, no. 6: 2079.

Richter, Daniel deB, and Dan H. Yaalon. 2012. "'The Changing Model of Soil' Revisited." *Soil Science Society of America Journal* 76, no. 3: 766.

Robbins, Jim. 2013. "The Hidden World Under Our Feet." *New York Times*, May 11.

Robinson, D. A., I. Fraser, and E. J. Dominati et al. 2014. "On the Value of Soil Resources in the Context of Natural Capital and Ecosystem Service Delivery." *Soil Science Society of America Journal 78*, no. 3: 685.

Rodaway, Paul. 1994. *Sensuous Geographies: Body, Sense, and Place.* New York: Routledge.

Rose, Hilary. 1983. "Hand, Brain, and Heart: A Feminist Epistemology for the Natural Sciences." *Signs: Journal of Women in Culture and Society 9*, no. 1.

———.1994. *Love, Power, and Knowledge: Towards a Feminist Transformation of the Sciences.* Cambridge: Polity Press.

———. 1996. "My Enemy's Enemy Is—Only Perhaps—My Friend." *Social Text 14* (46/57): 61–80.

Rose, Hilary, and Steven Rose. 1976. *The Radicalisation of Science: Ideology Of/In the Natural Sciences.* London: Macmillan.

Rose, Nikolas. 2007. *The Politics of Life Itself: Biomedecine, Power, and Subjectivity in the Twenty-First Century.* Princeton: Princeton University Press.

Roy, Arundhati. 2003. "Confronting Empire," speech given at the 2003 World Social Forum. Porto Alegre, ZNet https://zcomm.org/znetarticle/confronting-empire-by-arundhati-roy/.

Sánchez, Pedro A. 2004. "The Next Green Revolution." *New York Times*, October 6.

———. 2010. "Tripling Crop Yields in Tropical Africa." *Nature Geoscience 3*, no. 5: 299–300.

Sánchez Criado, Tomás, Israel Rodríguez-Giralt, and Arianna Mencaroni. 2016. "Care in the (Critical) Making: Open Prototyping, or the Radicalization of Independent-Living Politics." *ALTER, European Journal of Disability Research 10*, no. 1: 24–39.

Sandoval, C. 1991. "U.S. Third World Women: The Theory and Method of Oppositional Consciousness in the Postmodern World." *GENDERS 10*: 1–24.

———. 1995. "New Sciences: Cyborg Feminism and the Methodology of the Oppressed." In *The Cyborg Handbook.* Edited by Chris Hables Gray. New York: Routledge.

Satava, Richard M. 2004. "Telemedicine, Virtual Reality, and Other Technologies That Will Transform How Healthcare Is Provided," http://depts.washington.edu/biointel/future-of-healthcare-Tokyo-0412.doc.

Savransky, Martin. 2012. "An Ecology of Times: Modern Knowledge, Non-Modern Temporalities." In *Movements in Time: Revolution, Social Justice, and Times of Change.* Edited by C. Lawrence and N. Churn. Newcastle upon Tyne: Cambridge Scholars Publishing.

———.2014. "Wondering about What Matters: The Adventure of Relevance and a Social Science to Come." Paper presented at "Beyond Matter, Beyond Concerns?" organized by the STS-Barcelona group, Universitat Oberta de Catalunya, Barcelona, June 25.

Schiebinger, Londa, ed. 2003. "Gender and Science." *Signs: Journal of Women in Culture and Society* 28, no. 3.

Schmidtz, David. 2006. *The Elements of Justice.* Cambridge: Cambridge University Press.

Schrader, Astrid. 2010. "Responding to *Pfiesteria piscicida* (the Fish Killer): Phantomatic Ontologies, Indeterminacy, and Responsibility in Toxic Microbiology." *Social Studies of Science* 40, no. 2: 275–306.

———. 2012. "The Time of Slime: Anthropocentrism in Harmful Algal Research." *Environmental Philosophy* 9, no. 1: 71–94.

———. 2015. "Abyssal Intimacies and Temporalities of Care: How (Not) to Care about Deformed Leaf Bugs in the Aftermath of Chernobyl." *Social Studies of Science* 45, no. 5: 665–90.

Schriempf, Alexa. 2009. "Hearing Deafness: Subjectness, Articulateness, and Communicability." *Subjectivity* 28.

Sedgwick, Eve Kosofsky. 2003. *Touching Feeling: Affect, Pedagogy, Performativity.* Durham: Duke University Press.

Segall, Shlomi. 2005. "Unconditional Welfare Benefits and the Principle of Reciprocity." *Politics, Philosophy, Economics* 4, no. 3: 331–54.

Sevenhuijsen, Selma. 1998. *Citizenship and the Ethics of Care: Feminist Considerations on Justice, Morality, and Politics.* New York: Routledge.

Shapin, Steven, and Simon Schaffer. 1985. *Leviathan and the Air-Pump: Hobbes, Boyle, and the Experimental Life.* Princeton: Princeton University Press.

Shildrick, Margrit, and Roxanne Mykitiuk, eds. 2005. *Ethics of the Body: Postconventional Challenges.* Cambridge, Mass.: MIT Press.

Shiva, Vandana. 1991. *The Violence of Green Revolution: Third World Agriculture, Ecology, and Politics*: London: Zed Books.

———. 2008. *Soil Not Oil: Environmental Justice in a Time of Climate Crisis.* Cambridge, Mass.: South End Press.

Shiva, Vandana, Ingunn Moser, and Network Third World. 1995. *Biopolitics: A Feminist and Ecological Reader on Biotechnology.* Atlantic Highlands, N.J.: Penang.

Singleton, Vicky, and John Law. 2013. "Devices as Rituals: Notes on Enacting Resistance." *Journal of Cultural Economy* 6, no. 3: 259–77.

Sinha, R. K., D. Valani, V. Chandran, and B. K. Soni. 2011. *Earthworms, The Soil Managers: Their Role in Restoration and Improvement of Soil Fertility.* Hauppauge, N.Y.: Nova Science Publishers.

Sismondo, Sergio. 2008. "Science and Technology Studies and an Engaged Program." In *The Handbook of Science and Technology Studies.* Edited by Olga Amsterdamska, Michael Lynch, and Judy Wajcman, 13–32. Cambridge, Mass.: MIT Press.

———. 2010. *Introduction to Science and Technology Studies.* Malden, Mass.: Wiley-Blackwell.

Smith, Dorothy E. 1987. *The Everyday World as Problematic: A Feminist Sociology.* Boston: Northeastern University Press.

Sobchack, Vivian. 2004. *Carnal Thoughts: Embodiment and Moving Image Culture.* Berkeley: University of California Press.

Star, Susan Leigh. 1991. "Power, Technologies, and the Phenomenology of Conventions: On Being Allergic to Onions." In *A Sociology of Monsters: Essays on Power, Technology, and Domination.* Edited by John Law, 26–56. New York: Routledge.

———, ed. 1995. *Ecologies of Knowledge: Work and Politics in Science and Technology.* Albany: State University of New York Press.

———. 1999. "The Ethnography of Infrastructure." *American Behavioral Scientist* 43, no. 3: 377–91.

———. 2007. "Interview." In *Philosophy of Technology.* Edited by Jan Kyrre-Berg Olsen and Evan Selinger. New York: Automatic Press.

Star, Susan Leigh, and Geoffrey C. Bowker. 2007. "Enacting Silence: Residual Categories as a Challenge for Ethics, Information Systems, and Communication." *Ethics and Information Technology* 9, no. 4: 273–80.

Starhawk. 1987. *Truth or Dare: Encounters with Power, Authority, and Mystery.* San Francisco: Harper & Row.

———. 1999. *The Spiral Dance: A Rebirth of the Ancient Religion of the Great Goddess. 20th anniversary edition, with new introduction and chapter-by-chapter commentary.* San Francisco: HarperSanFrancisco.

———. 2002. *Webs of Power: Notes from the Global Uprising.* Gabriola Island, B.C.: New Society.

———. 2004. *The Earth Path: Grounding Your Spirit in the Rhythms of Nature.* San Francisco: HarperCollins.

Stengers, Isabelle. 1993. *L'invention des Sciences Modernes.* Paris: La Découverte.

———. 1997. *Cosmopolitiques.* Paris: La Découverte/Les empêcheurs de penser en rond.

———. 2000. *The Invention of Modern Science.* Minneapolis: University of Minnesota Press.

———. 2004. "Devenir philosophe: Un goût pour l'aventure?" In *La vocation philosophique.* Edited by Bayard-Centre Pompidou. Paris: Centre Pompidou-Bayard.

———. 2005. "The Cosmopolitical Proposal." In *Making Things Public: Atmospheres of Democracy.* Edited by Bruno Latour and Peter Weibel, 994–1003. Cambridge Mass.: MIT Press.

———. 2008. "Experimenting with Refrains: Subjectivity and the Challenge of Escaping Modern Dualism." *Subjectivity* 22.

———. 2010. *Cosmopolitics I.* Minneapolis: University of Minnesota Press.

———. 2012. "Reclaiming Animism." *e-flux* 36.

Stephenson, Neal. 1996. "Mother Earth, Motherboard." *Wired* 4, no. 12 (December): 97–160.

Stephenson, Niamh, and Dimitris Papadopoulos. 2006. *Analysing Everyday Experience: Social Research and Political Change.* London: Palgrave Macmillan.

Strand, Ginger. 2008. "Keyword: Evil; Google's Addiction to Cheap Electricity." *Harper's Magazine,* March.

Strathern, Marilyn. 2004. *Partial Connections*, updated ed. Walnut Creek, Calif.: AltaMira Press.

Stuart, Murray J. 2007. "Care and the Self: Biotechnology, Reproduction, and the Good Life." *Philosophy, Ethics, and Humanities in Medicine* 2, no. 6.

Stuart, Murray J., and Dave Holmes. 2009. *Critical Interventions in the Ethics of Healthcare: Challenging the Principle of Autonomy in Bioethics.* London: Ashgate.

Suchman, Lucy. 2007a. "Feminist STS and the Sciences of the Artificial." In *New Handbook of Science and Technology Studies.* Cambridge, Mass.: MIT Press.

———. 2007b. *Human-Machine Reconfigurations: Plans and Situated Actions,* 2nd ed. Cambridge: Cambridge University Press.

———. 2015. "Situational Awareness: Deadly Bioconvergence at the Boundaries of Bodies and Machines." *Media Tropes* 5, no. 1: 1–24.

Suchman, Lucy, and Libby Bishop. 2000. "Problematizing 'Innovation' as a Critical Project." *Technology Analysis & Strategic Management* 12, no. 3: 327–33.

Sunder Rajan, Kaushik. 2006. *Biocapital: The Constitution of Postgenomic Life.* Durham: Duke University Press.

———. 2007. "Experimental Values: Indian Clinical Trials and Surplus Health." *New Left Review* 45: 67–88.

Suzuki, Wakana. 2015. "The Care of the Cell." *NatureCulture,* no. 3: 87–105.

Swift, Mike. 2001. "Foreword." In *Soil Ecology.* Edited by Patrick Lavelle and Alister V. Spain, xix–xx. New York: Kluwer Academic.

Thompson, Charis. 2005. *Making Parents: The Ontological Choreography of Reproductive Technologies.* Cambridge, Mass.: MIT Press.

Thompson, Paul B. 1995. *The Spirit of the Soil: Agriculture and Environmental Ethics.* New York: Routledge.

———, ed. 2008. *The Ethics of Intensification: Agricultural Development and Cultural Change.* Dordrecht: Springer Science+Business Media.

Ticktin, Miriam. 2011. *Casualties of Care.* Berkeley: University of California Press.

Tomich, Thomas P., Sonja Brodt, and Howard Ferris et al. 2011. "Agroecology: A Review from a Global-Change Perspective." *Annual Review of Environment and Resources* 36, no. 1: 193–222.

Tomlinson, Isobel. 2011. "Doubling Food Production to Feed the 9 Billion: A Critical Perspective on a Key Discourse of Food Security in the UK." *Journal of Rural Studies* 29: 91–90.

Tronto, Joan C. 1987. "Beyond Gender Difference to a Theory of Care." *Signs: Journal of Women in Culture and Society* 12, no. 4: 644–63.

———. 1993. *Moral Boundaries: A Political Argument for an Ethic of Care.* New York: Routledge.

Tsiafouli, M. A., E. Thebault, and S. P. Sgardelis et al. 2014. "Intensive Agriculture Reduces Soil Biodiversity across Europe." *Glob. Chang. Biol.* 21, no. 2: 973–85.

Van Dooren, Thom. 2005. "'I would rather be a god/dess than a cyborg': A Pagan Encounter with Donna Haraway." *Pomegranate* 7, no. 1: 42–58.

———. 2014. *Flight Ways: Life and Loss at the Edge of Extinction*. New York: Columbia University Press.

van Leeuwen, Jeroen P., Lia Hemerik, Jaap Bloem, and Peter C. de Ruiter. 2011. "Food Webs and Ecosystem Services during Soil Transformations." *Applied Geochemistry* 26: S142.

Viseu, Ana. 2015. "Caring for Nanotechnology? Being an Integrated Social Scientist." *Social Studies of Science* 45, no. 5: 642–64.

Vora, Kalindi. 2009a. "The Commodification of Affect in Indian Call Centers." In *Intimate Labors: Interdisciplinary Perspectives on Care, Sex and Domestic Work*. Edited by Eileen Boris and Parrenas Rhacel. Stanford: Stanford University Press.

———. 2009b. "Indian Transnational Surrogacy and the Commodification of Vital Energy." *Subjectivity* 28: 266–78.

Wajcman, Judy. 2000. "Reflections on Gender and Technology Studies: In What State Is the Art?" *Social Studies of Science* 30, no. 3: 447–64.

Wardle, David A. 1999. "How Soil Food Webs Make Plants Grow." *Trends Ecol. Evol.* 14: 418–20.

———. 2002. *Communities and Ecosystems: Linking the Aboveground and Belowground Components*, vol. 34. Princeton: Princeton University Press.

Warkentin, Benno P. 1994. "Trends and Developments in Soil Science." In *The Literature of Soil Science*. Edited by P. McDonald. Ithaca: Cornell University Press.

———. 2006. *Footprints in the Soil: People and Ideas in Soil History*. Amsterdam: Elsevier.

Watson, Guy, and Jane Baxter. 2008. *Riverford Farm Cook Book*. London: Fourth Estate.

Weiss, Kenneth R. 2012. "In India, Agriculture's Green Revolution Dries Up." *Los Angeles Times*.

Whipp, Richard, Barbara Adam, and Ida Sabelis, eds. 2002. *Making Time: Time and Management in Modern Organizations*. Oxford: Oxford University Press.

Whitehead, Alfred North. 1920. *Concept of Nature*. Cambridge: Cambridge University Press.

Wild, Matthew 2010. "Peak Soil: It's Like Peak Oil, Only Worse." *Energy Bulletin*, May 13, http://dev.energybulletin.net/52788.

Wilkie, Alex, and Mike Michael. 2009. "Expectation and Mobilisation: Enacting Future Users." *Science, Technology & Human Values* 34, no. 4: 502–22.

Winner, Langdon. 1986. *The Whale and the Reactor: A Search for Limits in an Age of Technology*. Chicago: University of Chicago Press.

Wolfe, Cary. 2010. *What Is Posthumanism?* Minneapolis: University of Minnesota Press.

Woolf, Virginia. 1996. *A Room of One's Own and Three Guineas*. London: Vintage.

World Bank. 2013. "Growing Africa: Unlocking the Potential of Agribusiness." Washington, D.C.

Worster, Donald. 1979. *Dust Bowl: The Southern Plains in the 1930s*. New York: Oxford University Press.

Wyatt, S. 2007. "Review: Making Time and Taking Time." *Social Studies of Science* 37, no. 5: 821–24.

Yaalon, Dan H. 2000. "Soil Care Attitudes and Strategies of Land Use through Human History." *Sartoniana* 13: 147–59.

Yaalon, Dan H., and S. Berkowicz, eds. 1997. *History of Soil Science: International Perspectives.* Reiskirchen: Catena Verlag.

Yusoff, Kathryn. 2013. "Insensible Worlds: Postrelational Ethics, Indeterminacy, and the (K)nots of Relating." *Environment and Planning D: Society and Space* 31, no. 2: 208–26.

Zalasiewicz, J., M. Williams, A. Haywood, and M. Ellis. 2011. "The Anthropocene: A New Epoch of Geological Time?" *Philos. Trans. A Math. Phys. Eng. Sci.* 369 (1938): 835–41.

Zechmeister, Ingrid. 2001. "Foetal Images: The Power of Visual Technology in Antenatal Care and the Implications for Women's Reproductive Freedom." *Health Care Analysis* 9, no. 4: 387–400.

Index

(continued from page ii)